BREAST FEEDING AND SEXUALITY

Fertility, Reproduction and Sexuality

GENERAL EDITORS

David Parkin, *Director of the Institute of Social and Cultural Anthropology, University of Oxford*

Soraya Tremayne, *Co-ordinating Director of the Fertility and Reproduction Studies Group and Research Associate at the Institute of Social and Cultural Anthropology, University of Oxford and a Vice-President of the Royal Anthropological Institute*

BREAST FEEDING AND SEXUALITY

Behaviour, Beliefs and Taboos among the
Gogo Mothers in Tanzania

Mara Mabilia

Translated by
Mary S. Ash

Berghahn Books
New York • Oxford

First published in 2005 by

Berghahn Books

www.berghahnbooks.com

Library of Congress Cataloguing-in-Publication Data
Mabilia, Mara.
Breast feeding and sexuality : behaviour, beliefs and taboos
among the Gogo mothers in Tanzania / Mara Mabilia ; translated
by Mary S. Ash
    p. cm. -- (Fertility, reproduction and sexuality ; v. 5)
Includes bibliographical references and index.
ISBN 1-57181-677-1
1. Women. Gogo--Psychology. 2. Women, Gogo--Sexual
behaviour. 3. Women, Gogo--Attitudes. 4. Breastfeeding--
Tanzania--Cigongwe. 5. Sex Role--Tanzania--Cigongwe.
6. Taboo--Tanzania--Cigongwe. 7. Puerperium--Tanzania--
Cigongwe. 8. Cigongwe (Tanzania)--Social life and customs.
I. Title. II. Series

DT443.3.G64M33 2005-04-28
649'.33'09678—dc22
200504639

British Library Cataloguing in Publication Data
A catalogue record for this book is available from the British Library

Printed in the United States on acid-free paper

To my parents and sister

CONTENTS

ACKNOWLEDGEMENTS

I am grateful to all the people of Cigongwe whose conversation, friendliness and assistance made my research possible. Above all my thanks go to Kapwani's and Mkongola's families, to Milika Mpangwa, Moleni Lulenga, Margareth Samamba, Paolina Malogo, Edina Mahede, and to the staff of the Dispensary. Special thanks go to my field assistant, Mabada L. Caritas, and the person who introduced her to me, Andrew F. Rusatsi.

Thanking all the people who kindly helped me, particular thanks go to CUAMM-Doctors with Africa, which gave me the opportunity to do the research, and the CUAMM doctors who helped me in Tanzania and in Italy, particularly Anna Maria Dal Lago, Laura Guarenti and Massimo Serventi. My most grateful thanks are to the director of CUAMM-Doctors with Africa, Luigi Mazzucato, for his great kindness and help with my project.

I am indebted to Pat Caplan, Vanessa Maher, Antonio Marazzi, David Parkin, Donatella Schmidt, Paul Spencer and Elizabeth Whitaker for having exchanged ideas and points of view on my work. I particularly thank Renate Barber, my tutor in Oxford at the Centre for Cross-Cultural Studies on Women, for having given me constructive suggestions during our weekly meetings, and Murray Last, whose openness I always appreciated during my stay at University College of London. I owe special thanks to Soraya Tremayne for her encouragement and esteem.

Finally I wish to thank my translator, Mary Ash, for her competence and ability to interpret my thoughts.

Last but not least, I would like to express my gratitude to my parents and my sister who have never withdrawn their support and whose presence I took with me during the years I spent in the field.

ILLUSTRATIONS

Figure 1 *The hilly landscape of the field area*

Figure 2 *A mother pounding grain in the traditional way*

Figure 3 A young mother breast feeding her baby

Figure 4 A village woman in front of her hut

Figure 5 An extended family: three generation

Figure 6 A mother supervised by her mother-in-law

Figure 7 *A mother relaxing with her baby*

Map 1 *Tanzania*

INTRODUCTION

An anthropologist incurs a number of debts during his journeys: he owes something to whoever supported him financially, to the thoughts of other anthropologists and to the many people who helped him both practically and with friendship in the countries he has visited[1]

This book is the result of anthropological research that began in April 1989 and continued through to August 1992. It was part of the paediatric project which CUAMM, Doctors with Africa, was implementing in the government hospital of Dodoma and in the whole of the District.[2] This health intervention was centred on the problem of malnutrition and was led by Italian paediatricians, assisted by local health personnel, both in the hospital and in the territory.

In the area of Dodoma, the nutritional status of children under five years of age did not differ from the average national statistics. Around 50–60 percent of children suffered from moderate malnutrition, while 6–7 percent were severely malnourished, some peaks reaching 8–9 percent (MCH Report 1989/90). The health staff had found that, in general, a child that had been breast fed grew well for the first six months of life. The first slowing down of the growth curve occurred around this period, with a gradual worsening of the child's physical conditions accompanied by various pathologies, until reaching marasma and *kwashiorkor.*

The paediatricians' experience in the field, even more than simple data, indicated the existence of complex problems, many of which lay outside medical competence. From the indigestible integrated food to the precarious hygiene conditions, from the frequency, intensity and duration of breast feeding to the rules governing weaning processes, from the qualities attributed to breast milk to breast feeding from one breast only, from the changing availability of the mother to her conception of her newborn's growth rate – all of these factors indicated the presence and action of dynamics which interacted *in some way* with both the baby's process of growth and medical intervention. The health workers were in fact, beginning to be aware of a 'shadow zone', not immediately perceivable, but which deviously emerged when all the attention was focused on saving a child's life and which affected their efforts.

The anthropological research was included in this problem area, to discover which cultural and social organisational elements interacted in the infant feeding methods and what were the consequences of these on a child's development and health. Therefore, the mother–baby dyad, during the entire breast-feeding period of a time span between 24 and 30 months, became my preferred reference point.

I had planned to enter into the research by degrees, bearing in mind just how much an event which is so closely linked to female physiology is, in reality, loaded with complex cultural and symbolic values. Did I not come from a cultural area which, for the last two decades, had given up breast feeding its children as a sign of conquest for the emancipation of women? Renouncing breast feeding, isn't it perhaps an open manifestation of that move away from woman's 'most manifest animalism' as compared to man, of which Simone de Beauvoir spoke in her book, *The Second Sex* (De Beauvoir, 1961)? Was there not a lack, on women's part, of a careful reflection on the ever-increasing *medicalisation* of their reproductive health, including breast feeding?[3]

The reading undertaken in preparation for my research included, as well as the works of Rigby and Thiele, who preceded me in the field among the Wagogo, and of other anthropologist experts of East Africa, works on breast feeding and a series of medical manuals which were to help me understand female physiology and the great advantages of breast milk when compared to any other nutrient in the first months of an infant's life.[4]

My reaction to this latter reading was of 'discomfort' and reductivity: a woman was seen above all, if not exclusively, as a nurturer. Her being the mother of other children, a wife and also a daughter – and in any case a person with her own marked individuality – was never articulated. I asked myself how all of these aspects, through which a human being interacts with more or less defined and articulated social, relational and symbolic spaces could remain *silent*. Furthermore, the costs that a woman pays in terms of health in those countries in which breast feeding is the *condicio sine qua non* of the survival of her offspring, were rarely considered. The international organisations, in their public health programmes aimed at promoting the health of individuals, continued – and continue – to regard women always and only as mothers (the last, in order of time, are programmes aimed at the prevention of transmission of the HIV-1 virus in pregnancy). For women and their health needs the programmes tend to use the term 'mother and child', a binomial in which the second element is highly favoured, while their interdependency is evaluated only by the quality of the care that a mother gives her offspring.

Such a viewpoint seems not to realise two basic things: firstly, that a long period of breast feeding signifies that the baby has a long period of access to the mother's breast, an access which is accompanied, often much earlier than six months, by mixed feeding which puts at risk or impoverishes the advantages of breast feeding; secondly, it seems that they do not ask themselves about the incidence of cultural, social and economic reasons which can undermine the advantages of breast feeding in consideration of the high infant mortality precisely in those countries where mother's milk is offered to the infant for such a long time. These two

aspects, furthermore, should be accompanied by more careful consideration of the health of women who pass from one pregnancy to another and
who have to expend a great deal of energy fulfilling the role models and
tasks that belong to their lifestyles, which has serious consequences for
their well-being. A subsistence economy which sees them as producers of
food for the family nucleus, and to which other daily tasks are added –
gathering firewood, collecting water, caring for the family and the children – together with the more general conditions of poverty, insecurity
and even violence, takes a hard toll from a physique which is often malnourished as well.

A woman's health, therefore, together with that of her child, is not only
the result of her reproductive health, but also of her productive health,
because, as for a man, her life is based on the weaving together of both levels. Not to consider this means not to understand, in its entirety and complexity, that group of risk factors present in the environment and in daily
life which mean, for example, that a woman, who lives in what we continue to call 'developing countries', has 300 times more probability of dying
while giving birth than a woman in the so-called north of the world.[5]

My contact with village reality was to lead me to study aspects in which
the mother-child dyad was very far from the 'biological niche' which most
of medical literature was proclaiming. My interest in the relationship
between the physiology of breast feeding and the behaviour of the nurturer found, from the biocultural point of view, fertile ground for reflection on the connection between 'nature' and 'culture', so mobile in the
story of humanity – a mobility charged with implications for social and
biological life and, at the same time, dense with meaning for the understanding of phenomena which see the human organism responding,
adapting itself to new conditions, to new life models.[6]

It was only on my return from the field, and thanks to the work of
Vanessa Maher (1992), that I found a viewpoint that supported my own
ideas and that would give content and depth to what I had felt in the
preparation phase and then experimented later in the field.

If, as Maher wrote (1992: 153) 'factors [such] as political and economic
insecurity, ill health and overwork of mothers, gender inequality and the
dangerous and unhygienic environment that goes with sheer poverty' are
aspects which put the infant's health at risk, the same beliefs about this
nutritional practice can sanction its lack of success.

When I think back to the articulation of my work, to my daily contacts
with the women, to the different tracks, more or less explicitly suggested
to me by the women themselves on focusing and understanding the
dynamics connected to ways and attitudes, to those beliefs responding to
that particular nutritional system which is breast feeding, I can see that
my work was assuming such a value as to make an aspect of human physiology an act of culture – an act of culture which, also among the Wagogo,
assumes an importance, an absolute priority over any other role,
expectancy or need of a woman.

To give the breast, to give one's own milk, to have a healthy baby were to project a woman into a series of beliefs about the quality of her milk, dependent on behaviour, choices, obligations and rules, aimed at defining not merely the correct growth of her child but also her correct behaviour as a nurturer and, finally, as a mother.

This would lead me to consider the observation of a series of post-partum taboos, the first of which was total sexual abstinence during the breast-feeding period, a condition for the good quality of the nurturer's milk and guarantee for the correct growth of the infant. However, as proof of the sociocultural values of which the alimentary model, centred on breast feeding (a peculiarity of the female body), avails itself, the quality of maternal milk did in fact act as a control over the woman's behaviour in her role as nurturer and mother, thereby controlling her sexual behaviour as a wife and, in the last analysis, as a woman.

In the light of these considerations, breast feeding could have been configured in the dynamics of gift, a concept which, from Marcel Mauss (1965) to today, represents a key point in the analysis of those 'social facts' which also call into play, with the network of 'primary sociality', that *aimance* (Caillé, 2001) that binds a mother to her child with affection and emotion. This viewpoint will project the mother–child dyad into a wider context, delineating, above all, the outline of a woman who is complex and carrier of several instances which must be carefully considered.

I was in the field thirty years after Rigby and if, apparently, there seemed to be no difference between my time and what the British anthropologist dealt with in his experience, in my ethnographic reality I was considering how those changes present *in nuce* in the 1960s had assumed precise values with regard to social control. Those rules belonging to the Gogo tradition, those rules which gave order through initiation and marriage to family relationships, upset by economic changes or events influencing traditional pastoral economy, which had forced the men to emigrate to nearby urban centres, or to live on the margins of an economy that had marked their vision of the world for ever, had found, or perhaps just accentuated, a channel of control which had in the female body a strong ground of confirmation and denial.

Dealing with the study of breast feeding in order to try to understand the problems of malnutrition in the very first years of life meant an articulated study of women's sexual behaviour, thereby proving that it is not possible to comprehend the complex and composite dynamics of breast feeding, without asking oneself about the different roles that individuals are called upon to assume during their existence, without an articulated knowledge of the basic models on which the survival of a community are founded and, last but not least, without a vision of *gender*.

Once again, the fascinating research of the confines between nature and culture and their possible interrelations would give rise to more questions than answers. Furthermore, for the paediatricians working in the field, although the anthropological research would answer some of their questions, at the same time it would place their intervention in a confrontation with the mothers' beliefs, behaviours and choices, which needed to be understood and taken into consideration, with the aim of searching for and experimenting communicative contents having *reciprocal sense* in a continual, not always easy and taken-for-granted, labour of mediation. Therefore, even the paediatricians would be faced with the problem of the *'otherness* of thought and word' in individuals who were looking to them to alleviate the suffering of their loved ones.

Notes

1. Bateson (1980), 'Forward' to *Naven*.
2. CUAMM, Doctors with Africa is a nongovernmental organization from Padua, Italy, which has been working for fifty years in health cooperation in Africa.
3. For Italy see the interesting work of Elizabeth Dixon Whitaker (1994, 2000), which in an historical prospective considers the cultural dynamics through which the body and its functions are conditioned, moulded. In the same light, I mention Scheper-Hughes and Lock's interesting article (1987) and I recall the work of Merchant (1988), which emphasises the processuality of the dynamics through which culture models the body and its functions. More generally, all that we perceive as 'natural' feeds on cultural conceptions, prospectives and interpretations.
4. These readings were to give rise to a work written by more than one author, aimed at creating an interdisciplinary bibliography on maternal breast feeding (Campus et al., 1998).
5. *Osservatorio Italiano* on Global Health 2004 and WHO 2002.
6. Some readings were particularly useful: Wilson (1978); Konner (1982); Anderson (1983); Barash (1986); Cohen (1989); Durham (1991). Others will be indicated throughout the work.

CHAPTER 1

CIGONGWE[1]

Towards the village

In the last three months I had driven through the district of Dodoma and visited more than eighty villages, looking for the one that would be *mine*. Now that I had chosen it, I was travelling, and not without some emotion, towards that area which was to become very familiar to me.[2]

I had already left the city of Dodoma behind me and the tarmac road leading to the west was wide and bordered on both sides by low houses with corrugated iron roofs. The space in front of me, just a few metres from the roadside, was bustling with activity. Men and women went backwards and forwards or stopped along the way, the former busy with various tasks or chatting amongst themselves and the latter sitting under the canopies facing the houses, talking or playing *bao*.[3] Several children ran after a cloth ball, like a swarm of bees returning to the hive, others raced each other pushing old tyres or tin cans bearing high-flown sponsor names – Coca-Cola, Nestlé, Marlboro – towards dusty finishing lines.

While I was distracted by what was happening around me, the car jolted as the road became a dusty track full of potholes and the houses gave way to thorny hedges, through which dry and sandy expanses of land could be seen, dotted here and there with bushes. Scattered mud huts predicted recent settlements close to the town. The hedges suddenly finished, and I was able to see an undulating plain extending in all directions, in the distance flanked by hills from which enormous boulders emerged, made by who knows what hands. The track now wound its way, alternating ups and downs with long straight stretches of in deep ruts which cautioned a reduction of speed. This was quite a busy road, in the northwest leading to Bahi and the distant Tabora and Singida.

I encountered some men and women along the way, coming from or going to the town, and every now and again a dust cloud formed in the distance, announcing the arrival of a vehicle. These encounters shrouded us in a blinding cloud of dust from which it was advisable to escape as quickly as possible. After a few kilometres, my attention was caught by another dust cloud, but bigger and slower, as it came towards me. As it reached me I realised that it was caused by a herd of cattle. The thin and tired animals were listless, almost indifferent as they avoided my car. The men who were taking the cattle to the town or moving them in the search for pasture, appeared ghostly, their features flattened by the sand. After many tired and monotonous hours of walking, however, their faces regained shape as they offered smiles and shouts of greeting.

When the town disappeared once and for all from the rear-view mirror, the road began to climb, becoming steeper and rougher. I knew that I had to go through a pass, after which the village would become visible, but I did not remember such an uneven and hilly road. As I got closer to the hills on my left, I could see the piled-up boulders and stones on which sun-dried bushes seemed to cling for their lives, and here and there black shadows seemed to indicate the entrance to obscure caves. On the right, the road was lined with thin scrub, the brambles now and then interrupted by paths beaten down by who knows what passage of feet.

On reaching the pass, I was surprised by a number of baboons racing among the rocks as though to grab the best seats from which to better observe my arrival. They were soon to become my most faithful spectators.

From the highest point of the track I stopped and glimpsed below the first scattered huts, with their typical rectangular structure, of the village of Cigongwe, and at the same moment the clinic's corrugated iron roof reflected the sun's rays in a blinding glare. From my high vantage point, I could follow the contours of the land, the intermittently open and hidden spaces, the light and shade of the vegetation in which extraordinarily green acacias and imposing baobabs rose up like giant guardians. A spider's web of paths on both sides of the road branched out through the low and thin vegetation – they disappeared and reappeared towards the huts settled between the ups and downs of that rough terrain, red and in places sandy, everywhere extremely arid.

It was October and rain was only expected for the end of November or the beginning of December and would last, with violent and irregular downpours, until March or April.[4] My arrival coincided, therefore, with the dry season, when everything was barren and dried up by the many months without rain, and men, animals and plants moved sluggishly in expectation. A few months later I would notice how much the rain changed the scenery – a rapid change capable of suddenly covering a terrain broken by drought with a carpet of white bellflowers and reviving the scrub land, of germinating the millet and sorghum planted at the first sign of rain and of stimulating the spontaneous flowering of a variety of herbs and grasses to vary the diet of animals and men alike.

Before starting off again, my eyes followed the road below me as it descended and ran in a straight line for a stretch, only to overcome a dry river bed and begin its ascent in a bright red line which cut across a green obstacle in its path – a long range of hills covered with thick vegetation, running from north to south, a barrier between the village and what, as I came to know later, was one of its four sections, Msembeta. As I was getting into the car, I glimpsed a line of women coming along one of the many lateral paths further below, carrying something long and large on their heads. On looking closer, I realised that they were carrying firewood which was heavy to carry home and had been collected at the cost of great fatigue.

I began the descent and after about one kilometre, turned left and stopped not far from the track, in front of the headquarters of the *Chama Cha Mapinduzi* (CCM).[5]

The village chief, also the head of the political party, was waiting for me. I had met him a few weeks previously when, together with a medical doctor, I had gone to the village clinic for my first introduction. At that time, I told him that I would like to spend some time in the community getting to know its customs and traditions, the way of life and above all, the daily life of the women and the way in which they cared for their children during breast feeding and weaning. The man had shown interest and advised me to return after a few weeks as he needed to inform the *mabalozi*, each one of which was spokesman and responsible for a group of ten homesteads.[6]

There were a number of men waiting for me with the village chief and I was introduced with wide smiles and handshakes. The introductions would naturally not be finished that day as many of the *mabalozi* were not present – encounters continued for another two weeks. Accompanied by the chief, they allowed me to visit the whole area, to meet the community authorities and to inform them of the reasons for my presence there. Once the introductions were over, I was free to move where I wanted as I was known, directly or indirectly, to the inhabitants. I was soon to understand, in fact, the efficiency of the communication model based on 'pass the word on'.

Meeting with the village

The initial introductory phase in the village, during which I went from one end of it to the other, allowed me to ascertain the breadth of the area for my research and, at the same time, to contact a large number of people, most of whom I would meet only occasionally during my stay.

The beginning of the inhabited area from the southeast is about thirty kilometres from Dodoma, while in the west it is extensive and wider, collocating the last settlements at a longer distance from the town. It was difficult for me to calculate distances, as I always left the car on the track or on a pathway, sometimes in a field, to continue on foot for the whole day. Only by following the track coasting the village to the northwest, was I able to count exactly the nine kilometres to the next settlement, Cigwe.

The 3,864 inhabitants of Cigongwe[7] occupied a large territory, with scattered houses and large uninhabited areas, part of which were alternately left idle or cultivated, at that time covered here and there by stubble from the previous harvest. The inhabitants belong to one of the four sections (*mtaa*, pl. *mitaa*) composing the village: Mwuti, Mkombola,

Mleme, Msembeta.[8] Each one extends at a different distance from the main road. Mwuti, in fact, occupies an area on both sides of the communicating road, joining up with Mkombola and Mleme towards the west; while Msembeta is found on the other side of the hill in the north, to the right of the road.

The clinic, the primary school and a water point which takes water to the town's piped system are situated in Mwuti at the beginning of the village when coming from Dodoma, so the inhabitants can reach these services from different distances, or decide not to use them at all. As people move around on foot, getting to the clinic for those who do not live in the immediate vicinity[9] is problematic, especially during the rainy season or in the agricultural season which follows it. The water source is used mostly by the inhabitants of Mwuti and only for washing themselves and their clothes; for drinking and cooking they prefer to use water collected by digging a hole (*isima*) in the gravelly river bed. The first time I saw the wide, sandy bed of the water course cutting the village longitudinally, I thought that it was an important river. In actual fact, during the three rainy seasons that I spent in the village, I saw running water in only a part of the bed, and for just a few hours in one single day.

Those first few weeks spent in the village were to remain engraved on my mind.

I was very careful to observe everything and to look for points of reference for when I moved around on my own. Big and small baobabs, acacias and other richly foliaged trees helped me to calculate the road covered and to find my direction. I noted everything with interest: the habitations which were expressions of the composition of the family nucleus, the objects which had been left in the yards in front of the homesteads and which were clues to a daily life that I was soon to learn about, and the kraal in the centre of the yards, testifying to the presence of animal herds. Although the survival of the Wagogo depends on the cultivation of different varieties of sorghum and millet, their system of values is deeply marked by pastoral activities.[10]

Rigby's research underlined well the way in which cattle allow the Wagogo to manipulate their social system, entrusted on the one hand to residential mobility, and on the other to a wide variety of ties of kinship and affinity. The fact that their survival depends above all on agricultural activity, which is subject to the caprice of a single and often elusive rainy season, is secondary to the social and economic potential offered by cattle, together with the important influence in the political and ritualistic field. Right from the beginning, by talking together with the men, I was able to ascertain the place of cattle in the life and ideology of the Wagogo.

In one afternoon of my first week of wandering in the village I reached, together with my guide, the *mtaa* of Mkombola. After having followed a track which zigzagged its way between acacias and baobabs, we arrived in the vicinity of a European-style house, the only one present in the village, together with the clinic. We were warmly welcomed by a *balozi* who, after the ritual greetings, took us to the boundary of the large corral. By inviting me to observe this large space, empty at the time, his intention was to point out how the ground inside the enclosure was higher than that surrounding it. In fact, the embankment demonstrated both the presence of a large number of cattle and therefore the wealth of the household, and the prestige deriving from the fact that the area had been occupied for a long period of time, as shown by the accumulation and successive sedimentation of animal excrement. This was the reason why the men never failed to show me the animal enclosures in the yards in front of the houses, even though more often than not they were empty. During the dry season the animals do not return home every day, as the search for pastures forces the Wagogo to travel further and further away from the village, often for weeks. At that time, in fact, there were not many men present in the village, other than the *mabalozi* and those who were too old to take the cattle to pasture.[11]

Touring the village on foot forever accompanied by my host, I quickly had to get used to the presence of the children. They joined us along the road and their noisy chatter attracted others along the way. So as the day went on we were followed by a swarm of noisy and joyful children, ready to surround or overtake us, happy to be photographed at any time, assuming defiant, embarrassed or smiling poses in front of the camera lens. On reaching a hut they would run ahead, announcing our arrival with the racket they made. The adults welcomed us with never-ending greetings, while the youngest children ran away crying, frightened by our appearance or – as I was to discover time and time again – by *my* appearance. Once the ritual of greetings was over, we were offered stools to sit on just outside the hut and near to the entrance. After a while, the frightened little ones began to reappear and, still shy but won over by curiosity, one at a time they curled up in their mothers' laps.

It was always my host who began the conversation, by introducing me and then leaving it to me to explain the reasons for my presence there. It was the most delicate and, at the same time, the most exciting moment. It was my chance to make myself known and to meet the people with whom, within the next few days, I would be spending my time and on whom my research work depended. I felt as though I was on probation and I was afraid of making mistakes which would unknowingly prejudice future relationships. In those first encounters I tried to memorise names and faces, above all those of people who at first sight, due to that strange play of sensations which is triggered between human beings, seemed more willing, interested, vivacious and intuitive about what I was looking for.

They were very intense days, days in which I appreciated the people's hospitality and kindness, tasted their curiosity towards me, welcomed their sympathy and encouragement when I tried to use my initial and stumbling Cigogo, leaving behind my more fluent Kiswahili. I also learnt to accept their laughter and their making fun of me when I confused words, or when I used them in an incorrect, strange or ridiculous way. In those first contacts the Wagogo seemed cordial, eager to make a good impression and were particularly intrigued that a white person, a white *woman*, was willing to speak their language in order to learn about their community and daily life.

Talking with men and women, I told them I was in the village to study their traditions and customs. Above all, I was interested in talking with the women about their lives; how they obtained food for the family, how they cooked it, how they brought up their children. Only as the conversations continued did I bring up the subject of breast feeding and weaning. If I had announced these subjects as my main objective at the beginning of my talk with them, according to their identification models, they would have immediately and automatically collocated me in the medical field.[12] Two types of obstacles would have been created. Firstly, I would have been given the role of healer and this would have given rise to many expectations in people, expectations that I would not have been capable of satisfying. Secondly, this role would falsify the conversation, conditioning people to want to make a good impression, and therefore supplying answers which were in some way 'correct' and 'appropriate' *for* a doctor.

I also perceived that my host's presence, as village head and member of the political party, gave our meetings an official tone which very often hindered people's spontaneity. I was instinctively more and more aware of being the object of a performance in which my counterpart wished to favourably impress the outsider, according to officially sanctioned rules. I realised that it was very difficult for me to subtract myself from these dynamics. When my presence was repeated daily and without institutional filters, this attitude gradually lessened and finally disappeared, leaving other problems in its place, however, as we shall soon see. I certainly learned in that period what it meant to be the centre of attention, to be observed with interest and curiosity, and raising questions which I believe were a novelty for my interlocutors. I was not surprised (was I any less interested or curious?), and *the way* I answered their questions and their attentions could influence the following encounters and even my presence within the community.

What I had read before going into the field about agropastoral populations, the various socioeconomic, historical and political aspects of Tanzania, and even about the Wagogo themselves, was of no help to me at all in that moment. What I was experiencing with them was entrusted, first of all, to a way of being men and women marked by different routes, the sensations and the expectations of which belonged to the personal, *to oneself,* while at the same time reciprocally influencing each other. We were *all* actors in an interactive process marked by reciprocal perceptions and interactions of *the other,* by sentiments, appearances, experiences and interests capable of continuously composing and recomposing, rarely defined. The awareness of being able only partially to grasp the reactions I was creating by just my presence, of being able to evaluate the effects of my actions, for example by accepting their hospitality, brought me back to the words of Geertz (1974) on the controversial and complex need to understand 'how the natives think and perceive', when he writes that this is the central problem of Malinowski's diary (1967) – a real and profound problem, in which the anthropologist's professional training and his personal and intimate abilities, peculiar to each individual, are unavoidably intertwined.

During those first visits, I entrusted myself purely and simply to the immediacy of those face-to-face encounters, to my perception, to *my* personal way of relating with others, aware of the importance of being accepted and considered an individual with whom it could be stimulating to talk. It is possible that giving importance to *my* desire 'to learn about' the Wagogo in those initial conversations was one of the keys to my being acceptable to them and contributed to the definition of the network of relations on which I depended during my research.

Fieldwork

The language: communication problems

Once the official presentation period was over, I needed to find an interpreter to assist me, for at least the initial fieldwork phase, and at the same time to teach me Cigogo. Neither of these needs found an easy solution in the face of the problem of 'reciprocal perceptibility'.

As my aim was to study the behaviour and rules linked to breast feeding and weaning, I was looking for a woman, without doubt more suited than a man to accompany me in my fieldwork. She would have to know English well enough to initially bridge the gap between Kiswahili, of which I had a fair knowledge but not sufficient to launch myself into articulate conversations about my reasons for being in the village and to understand my assistant's replies, and Cigogo, which would become the principal language of my fieldwork. This necessity meant that I was looking for a young woman with a high school diploma.

After my first meeting with a young girl who had just finished secondary school, I realised that using English as a mediating language was problematic for many reasons. The girls who had attended secondary school in the town had only an elementary knowledge of English and, as I discovered, only an elementary or almost nonexistent knowledge of Cigogo. Nyerere's illuminated policy of making Kiswahili Tanzania's official language was relegating the traditional languages of many ethnic groups to a secondary level, impoverishing them, especially for the younger, urban generations.[13]

For these reasons, my first visit to the village was to prove a disaster. Other than the usual greetings which even I knew, and contrary to what she had previously said, my assistant did not know Cigogo. Another complication emerged with the second girl during our first visit to the village together. She had just finished secondary school, lived in Dodoma and was originally from Cigongwe. I thought that this would have facilitated any contacts, particularly with her relatives who were still resident there. After the first experience, I realised that using only English would limit the girl's comprehension and reduce the possibility of my learning both the local languages. I therefore dedicated an afternoon to explaining, half in English and half in Kiswahili, what was needed of her and what my intentions were. The morning after we left for the village.

We left the car at the clinic and began visiting the huts we found along the way. On meeting people, and especially the women, I saw that my assistant became increasingly uneasy and hesitant about going into the huts. I put it down to shyness. When we arrived by pure chance at a fairly large compound her uneasiness increased. Embarrassed, she asked me to wait outside while she went ahead alone to inquire if there was anyone around to talk to. Surprised at her request, I nevertheless waited outside, but felt as though I was being observed. And in fact, a number of women hurriedly appeared in doorways, only to disappear through others with furtive looks in my direction. When the girl returned she said that everyone was busy and without giving me time to say a word, began to move away. The day was nearly over and we set off for home, exchanging only a few words. We both had something to think about.

The next day, the girl did not appear for a return visit to the village, and I went to look for her at her home. It seemed that she was not present and a young girl told me that her sister did not want to work with me anymore.

As I made my way disconsolately back to the car, I met a woman whom I recognised as one of those who had 'spied' on me the day before in front of the large compound. After the ritual greetings, and on discovering that the young girl was her niece, I did not allow her to go on her way as she wished, and asked her to explain the girl's behaviour. Her answer somewhat disconcerted me: 'It's not dignified for a young person who has studied, to return to the village and to do a job which means asking questions left, right and centre, chatting about women's things'. This was the core of her answer, much more articulate and in lively Kiswahili.

So, while I thought I was offering a good job, my counterpart did not consider it either satisfactory or suitable for a young person with a diploma; on the contrary, it was embarrassing and even shameful for a young single woman. So much so as to force her to carefully avoid the involvement of her relatives living in the village and, in the end, to leave the job. The fact that the work was adequately paid had in no way influenced her final decision, which I was sure had been taken together with the family group.

This first encounter with the local reality was interesting and explanatory. First of all, and as further contacts with other young girls during my stay in Dodoma were to confirm, young people of both sexes who had had formal education hoped for work in the town and for a girl, preferably in an office.

For the new urbanised and educated generations, village life represented a return to the traditional past, to be escaped from once again. It did, however, still influence their daily behaviour. Proof of this was the reluctance on the part of young girls, not yet wives or mothers, to talk about maternity and breast feeding with adults. At the same time, women who were already married and had children considered it improper to talk about adult matters with young women. The different roles defined by generational differences still held sway, even with the young acculturised girls, proving the strong roots of traditional aspects.

Fate then lent me a hand. After a few more failed attempts, I was introduced to a forty-year-old woman, willing to accompany me to the village and to teach me Cigogo. This chance meeting would make a considerable contribution to my work and greatly influence the nature of my meetings in the village.

My new field assistant's knowledge of English facilitated our preliminary meetings, which were not limited to one afternoon. From the first, I perceived her interest in the study of specific aspects of the life of the Wagogo, where tradition and change continuously influenced each other, modifying the model of traditional life, above all in urban areas. The young women I had previously contacted had shown no interest at all in this aspect. She pointed out that by studying the village, we would be able to grasp traditional aspects still influencing or known to the older generation. This interest and love for her own people and for her own traditions which I sensed in her also manifested itself in her knowledge of Cigogo, which later proved to be excellent.

Knowledge of the language is truly one of the key aspects of fieldwork, not only because it allows dialogue without intermediaries, but also because it permits a more direct understanding of different models of expression resulting from the way of thinking and of presenting oneself, culturally typical of each human group. A firmer grasp of the language was to prove an indispensable instrument in deciphering double meanings, metaphors and metonymies, which are senseless when translated literally. The women, especially the older ones, made ample use of the above when explaining situations which were the subject of our conversations. The wealth of expressions and their use emphasised the way in which language presented itself according to the generation involved. Older people possessed a richness and complexity of expression which was lacking in the young. In the latter, Cigogo was less complex, almost meagre and more often than not mixed with Kiswahili. It was, however, easier for me to understand and at the beginning facilitated my learning. My assistant dealt with these complexities with extreme competence, while I was forced to return to childhood, learning the language by degrees, with patience, trying to overcome the embarrassment I felt when faced with interlocutors who loved to joke about my elementary attempts.

Although this created difficulties for me at the beginning, my assistant's help was providential, as she proved to be not just a simple interpreter but a *trait d'union* between me and the women. As I perceived it, she had an inborn talent for putting everyone at ease, with communication models which smoothly perceived the wealth and limits on both sides, and which reflected the best moments of her linguistic teachings. She taught me for the duration of my research and not only on a linguistic level. The fact that my assistant was a woman and a mother thwarted the embarrassment and the resistance that the women had previously shown towards the younger women. I myself was a problem in this respect, being neither wife nor mother. How was this overcome? How did I overcome their reluctance to talk about themselves to such a different person, so new to the categories used to define the *mzungu* (pl. *wazungu)*, a European?

My visibility

From the very beginning of this study, I believed that it was important not to overlook those factors linked to the personalities of individuals, to the way of perceiving 'the other' person and to 'the other' person's perception of me, to the quality of the relationships which would derive from this, to the variety of contacts which chance would repeatedly put my way, because, inevitably, all of this would lead to my theoretical orientation and to my choice of method. The field, therefore, should be experienced as a gymnasium in which the exercises, the relationships, model the *body* of knowledge, as well as the facts to be researched. Consideration of these aspects, outside the rules of 'scientific objectiveness', aids understanding of the anthropologist's efforts in the research and the evaluation of the facts to be studied. At the same time, the personal difficulties which may arise in the interaction with such a different human reality which is, at times, so irksome to meet, are taken into account. At this point, it becomes important to account for some of these aspects, because they became important, day after day, in tracing my *visibility* in the community. It was, furthermore, only by passing through the same, that the basis of what I was to discover and know was constructed.

My initial problem was to make the people I met understand that I was not a medical doctor. This was actually a relatively easy obstacle to overcome: my presence in the village, my choice of using a Gogo woman in aiding communication with the inhabitants, my interest in daily life, my declared desire to learn, all of these factors progressively, and rapidly, mitigated the idea of my being a *white healer*. It was, however, my interest in the female world above all, after an initial phase in which my attention was directed at both men and women, that was to change the way in which I was perceived by the community. Being with the women, as a woman, implied less formal behaviour, and becoming involved in household tasks facilitated increasing participation in the privacy of the wives and mothers.

And it is here that the *true* obstacle took shape: I was at an age by which Gogo women had already had more than one pregnancy and were surrounded by children, while I had none. In the first instance, the desire to enter their world seemed to be an unsuitable request, as my first assistants had been deemed unsuitable. It was not by chance that during my first meetings with them, and given my age, I never escaped the question as to whether or not I had children. After a few weeks this interest appeared to wane as, I was sure, 'word got round' the whole community, or at least the female part, of my *incompleteness*.

The women saw me as being different, as a *stranger* can be different, in the etymological sense of the term, *extra*, 'outside'. They knew that what was proper for the Gogo tradition, the two phases which turned a girl into a woman, ready for marriage and therefore for maternity – initiation and the rites of puberty – was not true for my group.[14] A woman is a woman, however, and as such, must have children. How did they manage then to ignore this deficiency of mine, a 'constitutional deficiency', and accept my presence among them, allowing me in some way to participate in their reality?

I think I can say that I was accepted initially because I was, above all, *anyway*, a woman and because of this they felt in some way pushed into 'sharing', into an attempt to have dialogue on that aspect which is not inborn and, at the same time, is so peculiar to the female physiology, breast feeding. I was a woman, even though *incomplete* according to their conceptions. The *wazungu* had, after all, introduced them to much stranger things! On the other hand, my difference was clear even when referred to their categories for 'whites'. They had met doctors, missionaries and nuns, whereas I presented myself to them as a person, a woman, desirous to know the reality of the Gogo woman and, in particular, her relationship with her child during breast feeding. It was a white person, a white woman to whom it was necessary to teach something about their daily lives, their own beliefs and traditions. This seemed to be a sufficient basis for the beginning of our relationship, in addition, I am sure, to a good dose of curiosity.

Although breast feeding, so ingrained in female physiology, was able to overcome my *inadequacy* as a woman, on its own it was not sufficient when I continued my research into the relations between men and women and, in particular, when I attempted to study in depth the rules structuring their sexual relations, an aspect which was to prove fundamental in understanding a mother's behaviour during the whole period devoted to breast feeding. At the same time, it was the women themselves who were to lead me onto the personal and intimate terrain that is sexuality.

It is not always easy to express in words when an event, or just a way of relating, a relationship, an encounter, signals that something new has happened or is taking place and changing the *sense* of the reciprocal presence, of the reciprocal consideration between individuals. My relations with the women were to change only gradually and over a period of months. It took time to appreciate fully the value of their availability towards me, which was the key to the turning point in their allowing me to participate in their feelings, in their way of being women at such an important moment in their lives.

My participation in daily life and my progress in Cigogo helped to define my presence, and above all to ratify my being a person, a person who was with them and interested in everything related to women and children. My *visibility* was confirmed after a few months by my being given an indigenous name, Matika, deriving from *itika*, the season of the year from February to April, when thanks to the rains, food is varied and abundant, a lucky name therefore. A year after my arrival in the village, one of the mothers named her child after me, rather ridiculous for the meaning that it has in Kiswahili, 'a time' [15], and unusual for the *r*, absent in the Gogo alphabet.[16]

It was in such moments that I felt fully accepted as a person, on the basis of my being a subject with whom 'it was possible to talk with each other'. How could I measure and evaluate the way in which the women saw me, if not through their way of relating with me? The way in which I related with them was certainly a desired result, strongly sought and obtained as well, according to indefinable lines of encounter that I had no control over, but which would just the same, determine the whole course of my research.

Even though I was conscious of the fact that there would always be an asymmetry between us, in as much as I was the one asking questions, seeking information and writing a book about them, *I myself* was often the object of their curiosity about the behaviour and beliefs of *my* group. Being, in my turn, the object of questions, allowed the relationship to run on a level of reciprocal recognition, of reciprocal *visibility*. To define my way of being with them as involvement in their lives is perhaps demanding and even pretentious, as I did not share their values, beliefs, conceptions on life, behavioural models and toil; but if by this term we intend to indicate an attempt to communicate with the objective of understanding, then there was involvement between us. From the exquisitely human point of view, with some of them there was also sharing, for which I am grateful.

The route of the research

The first approach

Familiarisation with village life began with my interest in domestic work, the preparation of meals, looking after children, agricultural work and husbandry. I wanted to get to know the community before going into what I was most interested in. I felt that it would be easier therefore, to begin with matters related to daily life, leaving my interlocutors free to lead me onto more private terrain, a reflection in any case of beliefs, behaviour, rules and prohibitions.

The vastly spread-out settlement of houses made me consider certain aspects of my surroundings so as to obtain a sample which would be homogeneously distributed throughout the territory: vicinity to or distance from the main communication roads, ease of access to water supplies, to the clinic and to the school. On choosing the different family nuclei I considered another two criteria: the presence or not of cattle and the presence of women with children of less than two years. In this way I selected 250 homesteads, little more than 25 percent of the total. I met with their inhabitants during what I was to consider the first phase of my research, which was to last about six months. I prepared three structured interviews to collect information about food production, daily nutrition and breast feeding and weaning, the first two addressed to both men and women and the last, for obvious reasons, to women only.

When I arrived at a homestead for the first time, I tried to keep the meeting on an informal level, to understand my interlocutors' interest and willingness to participate in my work. If the men and women agreed to answer my questions about their daily life, I began the interview or made an appointment with them within a few days. Before going into the various themes, I asked some questions aimed at collecting personal data: age, educational level, work and individual status, composition of the family and the occupation of the family head.

This ensured an informal start to our conversations, to put the interviewee at his ease. In fact, although I intended to follow a fixed scheme, I tried to carry out the interviews in a natural way, chatting and taking advantage of moments of apparent distraction, when neighbours or children arrived, to talk about fortuitous problems or events, no less interesting to note. This momentarily took attention away from what we were doing, but rendered the interaction more colloquial, thus helping the interviewee to overcome a certain degree of shyness due to the strangeness of the situation.

My presence and that of my assistant, answering questions and the use of the tape recorder all contrasted in fact, with the spontaneity which I wanted to achieve. The use of the recorder, for example, could have been a serious obstacle. But initial curiosity and some embarrassment about having to talk to 'a person holding a box', as I was initially called, were replaced by surprise and hilarity at hearing their own voices. These reactions soon allayed my fears about using the recorder, as it became for them a *thing* which allowed them to hear their own voices and to remember what they had said.[17]

Initially, my limited knowledge of the local idiom forced me to record all the interviews and most of the conversations, which were then written down by my assistant. This allowed me to read them together with her, and my fuller understanding of them improved my knowledge of Cigogo. Only when my linguistic knowledge improved did I fully appreciate the value of the chattering which often interrupted the interviews, the most valuable parts of which I noted down on my return home.

In general, however, even when I was linguistically more autonomous, I preferred to record the structured interviews and the meetings which followed in order to reflect on them later by listening to them several times. It was by listening often that I found the indications and the veiled suggestions for further study and for new directions to follow.

The part specifically dedicated to the study of the language, together with my assistant and through those recordings, represented an irremissible deepening of linguistic, and not only linguistic, learning. It took me closer to the *significant meaning* of words and expressions, to that 'imaginary universe' mentioned by Geertz (1987) which is so loaded with cultural contents, of their way of thinking and of facing life.

On passing from one family nucleus to another, constituting what I called my 'work group', it was soon clear to me that if the men agreed to be interviewed they generally preferred to answer immediately, and even when they agreed to a further meeting, they wanted to answer all the questions in one go. In any case, it was better to take advantage of their willingness and of their presence at home. It was the elderly men with more free time who, after a certain initial reserve or uneasiness perhaps in talking to a woman, dedicated me more time. Over the months, our conversations flowed freely, from their lives as young men defending the herds from wild animals and neighbours' raids, to the invasive presence of the Germans, and later, of the British. They were animated by speaking about their youth, and the continuous coming and going between past and present drew the outline of a reality which, for many reasons, they felt was elusive and inadequate for what had been taught them through the rites of passage and the long talks with their elders around the fire, after the tiring marches to bring the herds to new pastures.

With some of them in particular I formed a friendship and in our subsequent meetings they proved to be invaluable informants about the history of the village, the Gogo traditions and pastoral activities. Talking with these elders was not so much to do with my work plan, as with my involvement and acceptability, my pleasure, on a merely personal level, in talking with them, in *my* being the counterpart for once, in a relationship which I perceived started from others.

My encounters with the women were of a completely different nature. It could not have been otherwise as they were the protagonists of my study. Above all, it took more time for the women to answer a more articulated series of questions, and this meant returning more often to the same homestead, even remaining the whole day. This allowed me to observe what they were doing, in what order, how the tasks were divided between the family members, how much and what kind of attention they paid to the children, whether or not they were visited by neighbours or relatives and for what reasons, whether they were routine or occasional visits. In this way the interview was an excuse for starting a more long-lasting relationship with them.

In my attempt to become part of their daily lives, I often suggested following the women to the vegetable gardens near the house, to the fields at a few hours' walking distance, to collect water or wood for the kitchen. During these walks without my assistant, I asked them about their daily tasks and tried to improve my knowledge of the language by asking the names of everything I saw. The women satisfied my curiosity with the same patience and compliance which they used with the children. I could not help noticing the hilarity and, in some cases, irony, directed towards my ignorance as an adult. These were important moments, which increased our mutual familiarity and which also allowed me to 'savour' the kind of relationship which I had with different women, to 'feel' just how much my presence was daily becoming less of a novelty. Last, but not least, these conversations allowed me to weigh their interest in talking to me about the themes of my research.

The real problem

After the first six months, the knowledge gained and the relationship formed between myself and the villagers allowed me to go deeper into the themes of my research and therefore into closer contact with the women. The aspects of breast feeding which had emerged, where the women's ideas about the possible changes in the quality of breast milk had assumed a certain importance, gave a glimpse of a field of enquiry in which this nutritional technique, which the mothers called 'natural', in reality included a whole series of cultural and social indicators to be discovered and understood. It was therefore necessary to shift my attention from the mother–child dyad to the woman, her behaviour and her network of relationships during the period of breast feeding. The reason for this was that some women had, initially only by allusion and then more and more explicitly, cited sexual behaviour as an explanation for some of the most important modifications of mother's milk which could have serious effects on the child's health. In their allusions, their 'saying and not saying', I perceived that they were leading me, as wives and mothers, to the interpretation of their roles as such, both of which were equally and strongly influenced by learned beliefs, behaviour, rules and prohibitions and where fear, tension and suspicion could render a woman's family life difficult and cause her social relations to be interwoven with disapproval and outright censure.

On going deeper into these aspects of their lives, I understood that I needed a restricted number of mothers, meeting each one of them more frequently, thereby getting to know them, one by one, more intimately. If my daily presence, together with my linguistic progress, had granted me an ever-increasing *visibility* and presence in the life of the community, which meant more involvement both for myself and my hosts, it was probably no longer sufficient. In the choice of my interlocutors, I had to consider the willingness and the quality of their participation, together with the interest shown in answering my questions, as the involvement on the personal level would increase and for a longer period. The research programme foresaw a further 24 months in the field.

I realised that I was at a crucial point: this choice would, in one way or another, indicate the road to what I was searching for. I therefore reduced the number of mothers to 114, of which eighty-six were breast-feeding, while the remaining twenty-eight had begun mixed feeding, with some exceptions. Age varied, from young women between eighteen and twenty years, in their first motherhood, to adult women with more than one child. If these were the women that I intended to follow by observing their behaviour during their period of breast feeding and weaning, there were a dozen among them with whom my relationship was following the channels of true friendship, where my participation in their family life went beyond the bounds of the motives behind my research.

Giving my name to their last-born, asking my opinion in a discussion between women, receiving complaints about family life, being invited to a wedding during which the women sang about my participation in the event, and even the simple invitation to collect water – these all signalled my accepted presence and *visibility*, a *visibility* confirmed by an empathy which made them speak about themselves, their habits, beliefs, and fears about the health of their newborn. Their willingness to talk about themselves traced the outlines of their way of being wives and mothers during breast feeding, which, at the same time, I perceived as something problematic and complex.

Apart from this tight group of mothers, I could count on six women who, because of their age – most of them were no longer in the reproductive age group – and their role – two of them were traditional birth attendants and one of these a healer as well – had become precious informants for me. Their knowledge of Gogo traditions, together with their desire to make me participate, were a great help to my work.

As a starting point in this new phase of research I used an interview structured on the new points which had emerged and which required further examination, after the rather general one on breast feeding. The most profitable work was to come later when this systematically collected material would be studied more thoroughly. The opportunity might be represented by my visit, sometimes not planned, during which while chatting with the woman I would recall, through direct or indirect questions, aspects which had been emphasised by the same woman in a previous interview. Once I had recalled the woman's attention to the path I intended, I let her be the one to lead me along it.

These chats, only apparently informal, together with each woman's individual personality, incorporated a variety of reactions which went from silence to winking, from composure to jest. It was in this way that the women themselves spoke to me of post-partum taboos and of rules of behaviour between husband and wife and between each of them with other possible partners during the first months of the baby's life, thus rendering sexual behaviour one of the most important variables for the health of the breast-fed baby.

Conscious of the delicacy and complexity of the ground I was treading, I never tried to force my interlocutors in conversations for which they showed embarrassment or hesitation, returning in other moments to the themes which were important for me. I was not surprised by the women's various reactions. It is not easy to talk about the sphere of sexuality. On the one hand, it is represented by intimacy and deep and even conflicting emotions and sensations and, on the other, it is also the arena in which sociocultural plots and historical processes are measured.[18]

By my allowing the conversation to flow freely, each woman, when telling about one event and then another, one situation and then another, offered a cross-section of her own conceptions, convictions, predilections, apprehensions and principles. As well as noting her own personal way of dealing with the problems of breast feeding, I was able to perceive a con-

vergence on what the Gogo women themselves believed to be fundamental in their roles as wives and mothers.

In those moments I appreciated the consideration that the women
showed towards my person, even when they evaded my questions. Their
way of avoiding my requests could have been dictated by embarrassment
or bashfulness, or by the impropriety of speaking about *certain things* in
that moment or in that context. The presence of children or adolescents
was always a deterrent to 'adult conversation', while, at other times, the
presence of guests was a restraint to a dialogue which I had learnt was
more often than not an intimacy between them and myself. This made me
feel *accepted*. It was sufficient to be patient, it would go better the next
time. Sometimes the presence of a larger number of adult women helped
to conquer a certain bashfulness, on the basis of shared complicity, where
a joke or a butting in at the limits of good behaviour, or surpassing those
same limits, allowed embarrassing subjects to be dealt with, midway
between seriousness and jesting. It was thanks to these informal conversations that I was able to perceive aspects which were gradually acquiring
a central position. On this basis I reformulated other structured interviews
to be submitted to my sample.

Once the new round of interviews had been exhausted, I began my
informal coming and going among the women. This passage from the formal to the informal, and vice versa, allowed me, on the one hand, to deal
with complex problems methodically and, on the other, to obtain information on very intimate aspects of their lives as wives and mothers, about
which it was not easy to get answers.

The help of my group of female informants proved to be precious
indeed in this movement among the women. I did not hesitate in appeal-
ing to them for clarification, confirmation, indirect suggestions on how to
proceed and explanations of contradictions which, in my opinion, had
arisen from my conversations with one woman or another. It was an
opportunity to revisit certain aspects of their traditions and, at the same
time, to glimpse the changes in progress. At this point, I could not avoid
the humour and, at times, resentment of the older women on seeing the
way in which the life model that had shaped their infancy and part of
their adult life was being compromised. No less important, the collaboration of these women did not pass unobserved by the 'work group'.
Although each woman maintained the specificity accorded to each individual, and in thinking of each one I was able to trace the qualities peculiar to our relationship, the help that this group of older women offered
me, more or less consciously, exercised an important driving force on all
the others. With these women I realised the importance of learning
Cigogo in order to understand their way of telling and representing
themes such as those related to the sphere of sexuality, where the language was a highly condensed instrument, sometimes irksome, and even
mysterious.

I cannot but return, at the conclusion of this presentation of my work,
to the problem of communication. I was, in fact, covering ground in

which the understanding of the language, that 'sense of dialogue' mentioned by Geertz (1987), required not only a high level of specialisation, but also patience and humility.[19] This was the point of my research in which the role of my assistant became strategic. I was fortunate and grateful for the fact that she was able to unravel intricate expressive models, transforming a two-person dialogue into a choral one, aimed at gaining my participation in linguistic paths full of significance. These predicaments were the evident proof that my learning of the Gogo language was to last for the entire period of my work in the field.

Notes

1. In citing the toponyms, I use the Gogo writing which prefers the sound of the soft *c*, *Cigogo*, instead of the *ki* in Kiswahili.
2. The Dodoma District, altitude 1,000 metres above sea level, covers an area of 16,284 sq. km. Equal to 39.4 percent of the total of the region (41,311 sq. km) with 557,311 inhabitants – 353,478 in the rural area (14,004 sq. km) and 203,833 (2,280 sq. km) in the urban area (Census, 1988). The Wagogo are an agropastoral population of Bantu language and are the most numerous group in the District. The District occupies the southern part of the central highland of Tanzania, from the Rift Valley in the west to the Ruhebo mountains in the east, from the northern boundary of the Dodoma District to the Ruha River in the south. Today, the term *Ugogo* (the prefix *u* indicates the place, therefore the land of the Gogo) refers to the Dodoma District and the majority of the Mpwapwa District in the Dodoma Region, and a large part of the Manyoni District in the Singida Region.
3. A widespread game, with more or less important variations, in many sub-Saharan African countries, from east to west. It is played on a board (*bao*, in Kiswahili), having four rows of eight holes each, *mashimo* (sing. *shimo*), in which sixty-four seeds or stones, *kete*, are distributed. Each player has two rows with their respective seeds which must be moved so as to capture those of the adversary and to transfer them to one's own side of the board. The player who captures all his adversary's seeds is the winner (*How to Play Bao*, National Museum of Tanzania, 1971).
4. On average, the rainfall is 500–600 mm yearly. It varies, not only from year to year, but also from zone to zone. During my stay, rainfall was scarce in the area of the village. In fact, for the three rainy seasons, only one followed a normal pattern, while the other two were not even sufficient for the harvest, normally at survival limits anyway. The ecosystem demands great powers of adaptability from the Wagogo, starting from their mobility, in both agricultural production and husbandry. This mobility may not be sufficient when faced with famine due to lack of rainfall, to which other calamities – invasion by locusts and devastating epidemics (rinderpest) – may be added. Famines were frequent in the late nineteenth and twentieth centuries, and men and animals alike suffered greatly (Mnyampala, 1954; Iliffe, 1979; Thiele, 1984a, 1984b, 1985, 1986; Hadjivayanis 1989; Maddox, 1990, 1995).
5. *Chama Cha Mapinduzi* or Revolutionary Party. It was the result of the fusion, in 1977, between Julius Nyerere's party, the Tanganyika African National Union (TANU), founded in 1954, and the Afro-Shirazi Party (ASP) of Zanz-

ibar, constituted in 1956, under the guidance of Abaid Karume. At the time of my research TANU was still the only party in Tanzania.

6. In the *Ujamaa* structure, the *mabalozi* (sing. *balozi*) are the intermediaries between the inhabitants of the village and the Village Development Committee, the executive organ, made of up village members, which is responsible for the administration of the *Ujamaa* (Nyerere 1968; 1969).

7. Data from the last census. (United Republic of Tanzania, 1988) The total of the inhabitants resulting from the monthly supervisions to the clinic, carried out by the team, is higher by some hundred units.

8. On talking about the village, my elderly informants tell me that within the area indicated by the four toponyms they knew of others with their own names, and they always made reference to these when speaking of the Cigongwe territory.

9. I calculated that to reach the clinic in Mwuti on foot from the other *mitaa*, it took approximately: 30–60 minutes from Mkombola, from 50–80 from Mleme and 60–120 from Msembeta. I covered this distance without carrying weights of any kind.

10. East Africa's rich ethnographic literature, starting from the classic studies of Evans-Pritchard (1940), Gluckman (1950), Gulliver (1955), Beidelman (1960, 1961) and Le Vine (1964) to Rigby's study of the Wagogo in the early 1960s, has witnessed the 'cattle complex' which indicates cattle not only as an economic asset, but also as an asset imbued with emotional identification and as a tuning instrument for cultural attitudes, regulations and values (Rigby, 1969a; 1969b; 1971c). See Rigby's rich bibliography on the Wagogo to which a separate section has been dedicated. His works are repeatedly cited in the course of the present work. And, moreover, see Southon (1881); Cole (1902); Cluass, (1911); Hartnoll (1932); Schaegelen (1938a, b); Mnyampala (1954); Carnell (1955a, b); Gulliver (1959); Kwebena Nketia (1968).

11. Today, the absence of the men is not attributable only to husbandry. They move to the city, even as far as Morogoro or Dar es Salaam, or to other villages, in search of work. This necessity was evidenced by the presence in the village of hamlets without the traditional fencing for herds, while the data for the Dodoma District indicated that the total number of animals (cows, goats, sheep and donkeys) had been increasing for several years. The total of 6,069 head in 1987 had in fact increased to 6,653 in 1991. (This data was kindly given by the Dodoma District Council.) This total was not therefore uniformly distributed throughout the territory. Talking with the men in the village, I was able to understand that, to use their expression, 'cows go where there are cows', which means that the families already in possession of bovines were able to increase the size of their herds, while those who had lost their few, or many, animals due to raids, disease or having had to sell them when the harvest was absent or scarce, found it difficult or impossible to build up the herd again. Raids still go on in the area. More than two hundred head were stolen in one night, the wealth of several family units. They had been herded together for the night in a ravine not very far from the village, which was used as a resting place when moving to new pastures. The men who searched for them for days said that they must have been taken away by lorries, as they had disappeared completely.

12. The village inhabitants had two stereotypes for indicating a white woman: a nun, *sister*, or a doctor, *daktari*.

13. With the Education Act of 1978, the government made school attendance compulsory from seven to thirteen years of age; Universal Primary Education (UPE).

14. Among the Wagogo, rites of initiation are relative, as we shall see, to boys and girls. Only the latter are subject to puberty rites, which are carried out with the first menstrual period. This celebration is very important as it sanctions the girl's having reached marrying age, thereby enabling the definition of new affinity relationships (Rigby, 1967b).

15. *Mara*, noun and adverb: (1) a time, a single time, a turn, an occasion, an occurrence, (2) at once, immediately. A Standard Swahili–English Dictionary, Oxford University Press, first edition 1939 (1987). When I told them my name, I noticed the uncertainty in their expressions of those who are not sure they have understood correctly. So, to reassure them, I continued by saying: '*Mara, kama mara moja, mara mbili, mara nyingi* etc., Mara as at once, twice, often, repeatedly'.

16. See Johnston's study (1919); Cordell' grammar (1935); and 'Characteristics of the Gogo Group' in *Linguistic Survey of the Northern Bantu Borderland*, Oxford University Press for the International African Institute vol. IV (1956–7: 40–43).

17. One man defined my small recorder as *isumbi*, a hand-played musical instrument, which instead of turning off music, freed by the intervention of both thumbs, turned off the voice by the simple pressure of only one finger.

18. Some titles on sexual problems faced in anthropology are: Broude and Greene (1976); Strathern (1987); Cadwell et al. (1989, 1992); Le Blanc et al. (1991); Leavitt (1991); Lindebaum (1991); Ahlberg (1991, 1994); Tuzin (1991); Vance (1991) Heald (1995); Huygens et al. (1996). Other works will be cited throughout the work.

19. I like to remember Geertz's recollection of Stanley Cavel's thought on the difficulty of dialogue, 'a matter which is much more difficult than what is commonly assumed, and not only with foreigners' – dialogue which, for the scholar, is intended in the wide sense of the term and goes much further than just talking. Cavel's words recalled by Geertz should represent an a priori when starting a relationship with any individual: 'if to speak *for* someone else seems to be a mysterious procedure, it may be because to speak *to* someone else does not seem mysterious enough' (Geertz, 1987, Italian edition, p. 51). On the same interesting theme is the classic work by Victor Segalen (1907, Italian translation 1982).

CHAPTER 2

THE GOGO WOMEN

The road to understanding how Gogo women behave during breast feeding, that is, how they experience the nutritional process which occupies
them for more than two years, inevitably passes through their daily lives.
Here, their occupations, relations and ways of feeling all interact together
on breast feeding, conditioning how it is done in a composite network of
synergies. From this point of view, breast feeding is not a mere nutritional
'issue', but a complex exercise in which the mother–child couple is fully
integrated into social dynamics and is the bearer of cultural teachings.

It was therefore fundamental for me, in dealing with the theme of
breast feeding, to try to understand the female reality – made up of tasks,
commitments, bonds and pauses, but also of beliefs, sentiments, emotions
and expectations – through which the women shape their actions. I will
try to lead the reader along the road I took, above all to illustrate the scenario in which a subject of the female sex goes from being a child to an
adolescent, from an adolescent to a woman, wife and mother.

I glimpsed, in some aspects of the Gogo ethnography centred on space,
on the dwelling and on domestic objects, a way of approaching the female
world, in order to trace the outlines of the reality in which the women
model their behaviour and express their identity. It is not my intent therefore, along this road, to reach a complete cultural analysis of the Wagogo
environment and living space, but to grasp, through the same, some useful concepts for this work.

The living space

I discovered the full extent of the dwellings during my first movements:
rectangular huts with flat roofs which extended to an L-shaped or horseshoe form, according to the dimensions of the *kaya*, the traditionally
mobile Gogo residential unit.[1] It was from the number of doors, (sing.
mlango, pl. *milango*), that I was able to deduce the number of *nyumba*

(sing. and plural, house), the two rooms which form the house of a married woman with her children. I knew that the *nyumba* was composed of the *ikumbo*, the walls of which were not always plastered and which led both to the outside and to the internal room, the *kugati*.[2] This environment, as we shall see, is extremely important and significant for Gogo women, in their roles as wives and mothers.

The architecture of the living unit as seen from the outside, the shape and the number of doors, therefore gave me a fairly exact indication of the size of the residential nucleus. Furthermore, the presence of cattle was easily confirmed by simple observation.

In fact, when the ground in front of the *nyumba*, the *ibululu*, was closed off by a fence, by large poles or, more commonly, by a thick hedge of shrubs tied up in bundles, it usually contained another fenced-off area, *igagala*,[3] which housed the animals brought back from pasture. An opening, *ideha*, on the western side of the external barrier, allowed people and animals to enter.[4] I was to learn that this opening was closed at sunset with a gate or trellis of woven shrubs or with a simple bush of thorny branches, for the protection of the inhabitants. When there were no herds, there was no fence around the space in front of the *kaya*. On encountering these constructions, all more or less along the lines of the model described, I always tried to control their orientation with regard to the four cardinal points, as described by Rigby (1966a; 1969b).

Like Rigby, when he carried out his research, I saw only a few huts in the area in question that had not been built according to tradition, and that is, with the first dwelling, *nyumba ya cilima* (the house to the east), or *nyumba imbaha* (the oldest house; lit. the big house), placed on a north-to-south line, with the entrance door towards the west. The other *nyumba* are later added on to this as the residential unit develops, being attached to the former on the western side.

When I had the chance of observing a dwelling being moved, due to the scission of a residential group, the answers given to my questions about the building techniques confirmed the traditional values, which took me back to the spatial conceptuality previously pointed out by Rigby. 'My father and my father's father taught us not to cut the earth, *kudumula yisi*[5] [literally, to cut the country], and not to disturb the wind, *mbeho*', was the concise reply I received.

The wind, *mbeho*, is a mobile atmospheric element, the dynamics of which are perceivable on the skin and clearly visible when, blowing strongly, it creates whirlwinds here and there which carry sand and dry sticks and leaves around the village. What did it mean 'not to cut the earth' and 'not to disturb the wind'? Was it to explain that, although the wind carrying dust and sand arrives from the east, it is associated, in the Wagogo mind, with light, well-being and fertility, as opposed to darkness, death and illness associated with the west? Or that it is towards the west that illness and contamination are always 'thrown' in all the purification rites, whether referred to humans or to animals? The answer to the latter question was above all very practical: it is not possible to throw illness and contamination to the east; the wind blowing from there would carry it back. Throwing them to the west is like throwing them away twice, the wind will blow them further and further away.

This is a purely functional explanation; it is, however, only in the light of the Wagogo system of symbolic classification, that these ideas and values become comprehensible. This system is based on two fundamental series of oppositions, left/right and female/male, where emphasis is placed on the complementarity of these opposing tracks themselves, rather than on a relationship of inferiority to superiority.[6] It is indeed starting from these two pairs of oppositions that the Wagogo have redesigned and interpreted the different aspects of their spatiality (Rigby, 1966a; 1968b; 1969b).

The house must therefore be built with the correct orientation, where the four cardinal points are in oppositions and where, on one side, there is the east, the south, above, men, the right hand, light, fertility and life; and on the other, the west, the north, below, women, the left hand, darkness, sterility and death. Furthermore, the 'centre' must be added to the places of space designated by these cardinal points. Together these are perceived as exposure points, openings through which dangers for people, animals and harvests may enter.[7]

What role does the wind play in this configuration of space? If *mbeho* is the wind, an atmospheric phenomenon, it is understood, at the same time, as the 'ritual state' of a given area, which can be 'good', *mbeho swanu*, or 'bad', *mbeho ibi* or *ibeho*, according to the circumstances (Rigby, 1966a; 1968b). It is the latter acceptation of 'ritual state' which fulfils their idea of space and which receives more attention. I was quickly able to ascertain directly the complexity and multiple references of the term *mbeho* as a 'ritual state', as it remains deeply linked to events in the life of the Wagogo, from the rains to the harvests, from the building of a house to burial, from the opening of the initiation ceremonies to the rituals of fertility and protection.

I met up with this concept for the first time on the occasion of the move of a family group. A man, with his two wives, the children and the elderly mother, decided to move his home from the Mkombola area to Muwuti, before the rainy season.[8] This event allowed me to be present at the traditional lighting of the first fire with new sticks (*mhejo*, pl. *lupejeho*), obtained from the head of the *kaya*. After use, the sticks were buried in the kraal, together with special medicines supplied by the diviner. In this ceremony, the sticks are *mbeho*, and synthesise the ritual purity of the new dwelling, a purity directed to the protection of the occupants' health, of the herds and against invasion from wild animals.

In the second year of my stay in the village, the rains were late and very light. These circumstances were attributed to a bad 'ritual state', *ibeho*, caused by the lack of respect towards the spirits of the dead (*milungu*), when permission had been given for the work of an excavator near the slopes of mount Mlindimo, their recognised site.[9]

Just as the concept of *mbeho* is called into play for events which affect a group of people or the community, so it is equally considered in reference to the individual. We will return to this aspect extensively. I would just like to disclose here the words of one woman: '*So mbeho yo muwili hono visite, akutuga nhamwa yiyo*' (when the wind and the body refuse each other, the body becomes ill). With these words, she wanted to underline how an illness can be the result of the lack of equilibrium between an individual and the external environment (*mbeho*), the external social environment. The use of the verb *visite*, from the infinitive *kusita*, to refuse, accentuates, in this context, a condition which has been altered by an unusual event.

The Wagogo are aware that the 'ritual state' of an area is subject to natural and social events which may alternate favourable states with unfavourable ones (Rigby, 1968b). In the same way, favourable and unfavourable states may reflect on single individuals when the lack of respect of traditional rules, and not only those relative to the building of dwellings, upsets an equilibrium which should guarantee a physical state of well-being, of health. The alternation of states of equilibrium and non-equilibrium which affect the individual have specific meanings referred to the female body, and this will be one of the fundamental issues in understanding the ideas and the behaviour of mothers during the period of breast feeding.

The house

I had been anxious to go into a hut from my very first contacts with the population of Cigongwe. I knew about the structure from my reading and, maybe because of that, I could not wait to find myself inside one. I also believed that the invitation to enter was an important step towards the *visibility* that I was seeking.

In the first meetings, I was introduced to people in the courtyard in front of the houses, or near the fence opening. I was thus able to touch the distance separating us, and to remind myself of the use of space as a specific cultural elaboration, so clearly presented in Edward Hall's studies on proxemics.[10]

My wandering around between the huts and in the ample spaces which separated them, allowed me in any case to become acquainted with both the daily life of the village, with the coming and going of men and women, and with the 'courtyards', where the domestic objects left on the ground in front of the houses testified to the women's daily tasks. During the day, I observed those daily activities related to the preparation of food, or to the cleaning of the house and its external spaces, and therefore the women's expert use of those objects often left negligently on the *ibululu*. The mortar, *ituli*, obtained from a tree trunk and in the hollowed part of which cereals were placed to be separated from the husk, was ever present, together with a pestle, *mtwango* (pl. *mi-*), a sturdy stick about one metre long, the larger, rounded end of which was used to crush the seeds. The women and the young girls held it with both hands and vigorously and rhythmically lowered it into the mortar. As they lifted their arms, they stretched and bent their backs, adding energy to the thrust of their upper limbs. Sometimes the work was done in twos, offering me a demonstration of harmonious rhythmic ability.

I often saw the rectangular stone with rounded corners, used to reduce the seeds into flour, *luzala*. It was placed on two stones of different dimensions, thereby giving it a slope onto which the flour, obtained by friction, slid down into half a pumpkin, or onto an animal skin placed under the lower border of the surface. Given its reduced dimensions, the women can quite easily move it outside from the *ikumbo*, or vice versa. Just as for the *ituli*, the *luzala* called for dexterity and coordination, as one woman showed me. On her knees near the highest part of the stone, she placed a small quantity, more or less a handful, of millet on the sloping stone surface and began to rub the seeds with a rhythmic backwards and forwards movement, using an oval-shaped flat stone, *sago*, held with both hands. As the grains gradually broke up, she added more, allowing the fine flour to flow down onto a piece of ox skin. The whole operation took about half an hour and, as I was able to confirm by accepting the woman's invitation to try for myself, those gestures, which had seemed to be so rhythmically natural, were in fact the result of an acquired technique, probably learnt through imitation and which for a young girl at her first attempts must have required practice and diligence.

I also saw many other instruments: brooms similar to forks with long wooden prongs, *ikusililo*, or brushes made of a bundle of flexible sticks tied together with strips of vegetable fibre or skin, *iteyo*, with which the women swept the floor or the courtyard, just outside the house. Calabashes of various shapes and sizes were on the ground just outside the entrance. The variety in size and shape made me curious about their use. I learnt their specific names and the various techniques used for emptying and drying them, open or closed, according to whether they were to contain liquids or solids. Apart from the variety, I knew that these calabashes, *n'ghungu*, were symbolically linked to women, just as the bow, *upinde*, is to men (Rigby, 1966a).

When I arrived in the courtyard, the women welcomed me with the lengthy greetings, so typical of African tradition, accompanied by repeated handshakes and smiles, probably to encourage, at least at first, my shaky Cigogo. They invited me to sit on a stool, *igoda* (pl. *ma-*), promptly brought by a child who then left or crouched between its mother's legs, shyly or fearfully scrutinising me.

In these first rather informal encounters, they seated me near the door of the hut where, during the conversation, I could observe the partially plastered walls which revealed a thick structure of poles, approximately of the same size, placed vertically and horizontally, woven to two or three others, and fixed together halfway up the wall, which was little more than 1.5 or 1.6 metres.

There are no windows in the huts, but light and air enter through open fissures at more or less regular intervals in the well-plastered part of the building, normally the *kugati*, where the walls meet at the roof. These fissures, however, as I was soon able to ascertain, guaranteed only limited light to enter and limited air to circulate. The roof juts out thirty or forty centimetres, parallel to the ground. By looking up from my position seated on a stool, I was able to see the lines of poles, thicker than those used for the walls, spaced out by beams and held together with red clay, which formed the covering of the hut. Calabashes, clay pots and animal skins were on display on them, the latter used to hold vegetables laid out to dry and which, together with thick porridge made from millet and sorghum, *ugali*, constitute the main dish of their single daily meal. This solid covering rests on large external beams, supported by others, suitably spaced and perpendicularly driven into the walls. These design the outline of the dwelling, ensuring a robust external structure, a supplementary support frame.[11]

In harvest time, from May to June, another pole is added and placed on the jutting out part of the roof. Its branches, a good half metre from the roof, then display corn cobs left to dry, *msijiti wa watama*. The visual effect was singular; it looked like a strange bunch of fruit, normally not more than forty, all pointing straight up to the sky. During my stay, I noted the care with which they maintained the bearing structure, replacing the external supporting poles as soon as it was deemed necessary. The flat roof covering itself was also reinforced, or renewed, with clay before the rainy season, to maintain its thickness and solidity.

The sporadic but violent showers, typical of the area from November or December through to March or April, demand periodic maintenance of the inhabited structure which is entrusted, depending on the specific parts, to the men or the women. As for the building work, maintenance follows the traditional division of tasks between men and women. The former are expected to build the wooden structure, that is, the framework of the house, the covering and the maintenance of the roof (in which the women also participate by handing the dampened clay in baskets to the men on the roof). The women alone plaster the walls, firstly with a layer of grainy but resistant hard plaster, followed by a second layer using fine sand mixed with water which gives a smooth and compact surface. This work is done for both the internal and external parts of the *kugati*, while the *ikumbo* may be covered by a rough plaster which simply closes the spaces between one piece of wood and another, leaving the wooden framework largely visible. At night this room is used for the young animals, calves, sheep and goats and, quite often, as a sleeping area for adult males from the *nyumba*.[12]

By observing the maintenance and building work of a house, events which occasionally occurred during my first months in the village, I was able to observe just how the division of tasks between men and women was experienced within a complementary context, where any reference to superiority or inferiority of the tasks carried out was diluted by the climate of cooperation. The fact that the walls were considered female and the roof male, is a mere analogy with the position during sexual intercourse, and not a conception in terms of superiority and inferiority.[13]

One day, an elderly woman invited me to enter and watch her while she prepared a meal, and to share it with her. This, I felt, was a turning point.

Nyumba: physical-residential unit, social-residential unit

I remember my emotions when, for the first time, I passed over the threshold of a hut and found myself in the *ikumbo*, a bare room which was fairly luminous as the walls were not plastered. At first glance it appeared quite large: a series of three robust supporting poles, a couple of metres one from the other, were aligned in two parallel rows. I was immediately taken by their shape, as the parts which met at the roof level were heavy and shaped differently, squared or rounded. The room must have housed a large number of animals during the night, as the floor was covered by a slimy mud which, judging from the smell, was a mixture of sand, urine and dung.

My host, a very old woman, at least in appearance, a *bibi*, led me through the door to the left of the one through which I had entered, and into the *kugati*. I was forced to literally step over the threshold, which had a low wall of about twenty centimetres at its base, as though the door had been made by opening a space in the wall itself. Once over the threshold, I could see nothing in the darkness and I instinctively grasped the woman's arm. At the same time, the smoke lingering in the room burnt my eyes and throat, and the acrid smell was so strong it was to remain with me for a long time, even though I was to get used to it.

I tried to maintain an erect position by gaining a couple of centimetres between one beam and another and this was a strategy that I would always be forced to adopt. In fact, although on entering the houses, one was forced to step down about ten centimetres onto the internal floor, the height of the room could only just accommodate my 175 centimetres.

Still holding on to the woman's arm, I advanced into the gloom in which I was not able to distinguish clear contours. On reaching the middle of the room, I was able to make out the usual series of supporting poles blackened by the smoke of many fires. When my eyes got used to the darkness, I could also make out the usual variety of pumpkins and calabashes, a few earthenware saucepans and some stools all jumbled together on the floor, some near the wall on my right and some around the supporting poles. Other earthenware saucepans, blackened from use, were suspended from the ceiling in a plaited rope, *ikangambwa* (pl. *ma-*), in an orderly fashion from the biggest to the smallest. A dress and some *kanga*, the coloured cloths which the women wear tied around the hips or under the arms, were thrown over a string hung between the wall and a pole. A more attentive scrutiny also revealed the traditional piece of black cloth for which the Wagogo are famous. The women often use it in place of the coloured and more expensive *kanga*, while I only rarely saw it being used by the men, tied at the waist, or by just a few elderly men, tied to the right shoulder and draped along the body, leaving the knees uncovered.[14]

I glimpsed the bed, *ulili*, a suitably shaped elevation on the floor, covered with the skin of an ox, *nghin'go*.

The room must have been about five or six metres by three and a half, and in the furthest corner, I could just see the shape of a large basket raised about fifteen centimetres from the ground on large stones. This was the granary, *idong'ha* (pl. *ma-*), where the women keep the harvest of cereals, various qualities of millet and sorghum, *uhemba*,[15] the staple diet of the Wagogo. On my left, was the fireplace, three large rocks placed in a triangle, *mafigwa*. The embers, left over from the fire which had burnt throughout the night to ward off mosquitoes and other insects, gave out a soft light. A series of wooden spoons and ladles of different sizes were displayed on the wall above, strung one after the other between the wall and the rope, the ends of which had been hammered into the plaster.

My interest in what I was seeing – the house, its spaces, the objects – was alerted, as they would help me to understand some aspects of the reality of the Gogo women. All of this was introducing me to daily life made up of useful objects, habits, gestures and significance to be deciphered. My attention was captured not only by what I was seeing, but also by those lines, contents and conventions of customs unknown to me and which caused me to move with caution, not only because of the darkness.

The house therefore, in Suzanne Blier's words, came towards me as a 'living organism' (Blier, 1987: 2), with a story of its own, in which the relations between individuals, between individuals and space and between individuals and objects mix together, hide each other and reveal each other only if you know how to look.[16] I knew that the house was a crucial container in the life of the Gogo women. Not only because, as a physical-residential unit composed of two rooms, it is the area in which important activities take form and place – from the preparation of food to the care of children, from sexual activity to the birth of children – but also because it represents a social-relational unit, composed of a married woman and her offspring.

For a married woman, the *nyumba* is 'her' place above all others: it belongs to her entirely, so much so that she may keep the connecting door between the *ikumbo* and the *kugati* locked. No one can enter without her permission, not even her husband. It is in the *kugati* that the woman keeps her cooking utensils which, together with the fireplace, are the symbols of her dominion. It is in this space that her status, first as a wife and then as a mother, materialises and is defined. If these two rooms are invested with a strong symbolic value for a woman, married and a mother, they are no less so for the man, married and a father. If he does not build them, tension arises between the couple, tension which can lead to the dissolution of the marriage; and, should the man die before having carried out this duty, he is treated in the same way as a young man, having no right to a funeral ceremony.

In the past, when the presence of cattle was part of the economy of almost every *kaya*, the birth of the first son signified the attribution of a few head of cattle from the herd to the *nyumba*. This tradition is respected when the economy of the family nucleus can count on the cattle. In this case, the woman manages the products of the entrusted animals, from the milk which she herself milks, to the calves which remain part of the patrimony of the *nyumba* and which will be inherited only by her children. The shortage of animals and the impossibility of rebuilding the herd not only reduces the subsistence economy of the *kaya*, as the traditional means of barter for obtaining food in times of famine is also short, but weakens the woman's possibilities of working in favour of her children, thanks to the strategies presented by the cattle allocated to the *nyumba*.[17]

To summarise what has so far been said about the dwelling unit, the roles of men and women configure a male role undoubtedly associated with the *kaya*, and therefore with specific social categories, like the role of chief, the agnatic descent, wealth, and *in primis*, cattle when it is present, while the *nyumba* are clearly associated with the female figure. The woman, in her double role as wife and mother, continues to guarantee the continuity of the agnatic group, of the survival of the family group through reproduction, the care of the individual and the production and control of food.

If, on the one hand, the changes that have taken place, such as the unhomogeneous presence of herds in the *kaya*, have forced many males to seek other survival activities, on the other, they have equally accentuated the female role as guarantor of subsistence in an agricultural economy that is continuously uncertain and no longer guaranteed by cattle as instruments of barter when the harvest is lost or insufficient.[18]

The Gogo woman

The woman who had invited me was the mother of the chief of the *kaya*, in which there were two other adult women, her son's wives, with their respective sons and daughters. One of her grandsons had recently married and had already built the *itembe* for his young wife.[19] I was, therefore, in the *nyumba ya cilima* or *nyumba imbaha*, the house of the oldest woman of the *kaya*, in this case, the mother of the chief. I was in the *kugati*, where the grandson's young wife must have spent the first weeks after her wedding waiting for her new house, and where, as tradition demands, she must have been covered with an oily substance mixed with particular medicines, which ensures fertility and health to the new marriage. This oil is always conserved in the mother's house, or in the house of the chief's first wife, if the mother is no longer present.

Before my arrival, my host had already ground the millet and after having revived the fire with straw, twigs and pieces of wood, we were waiting for the water to boil to cook the porridge, *ugali*. In the meantime, she proudly told me that she was still able to tend her fields, even though it was becoming harder for her, as she was a *bibi*, an old lady.[20] Luckily, her fields were not far from the house and she could count on help from her daughters-in-law and grandchildren. Other women, she told me, had to walk three or four hours to reach their fields.

Agricultural activity, aimed at feeding the components of the *nyumba*, is mainly conducted by the women, although the men, according to the logic of the division of labour, clean and burn the fields, *kutemanga mbago*, while the hoeing is carried out together with the women. Another task, exclusively male by tradition, is the threshing of wheat, *kutowa uhemba*, but I very often saw women and children threshing the harvest near the huts with no contribution from the men.

Each woman has her own fields, *migunda*, where she grows the basic products necessary to satisfy her nutritional needs, those of her children, her husband and when necessary, her guests. She also has small vegetable plots, *vigunda*, usually near the house, where she grows maize, vegetables and groundnuts; she may sell the latter, if she wishes, at the nearby markets of Cigwe or Dodoma. The *nyumba* is therefore an entirely independent economic unit within the domestic group for the production, conservation, distribution and consumption of products, fruit of her agricultural labour.

I carefully watched the elderly woman prepare the meal and her gestures resembled the natural ability of a well-learned lesson. At the same time, I was attentive to what she was saying; her swift changes of subject and the wealth of her vocabulary, so typical of the older women, were a real challenge to my Cigogo. After cooking the thick porridge, which took only ten minutes as the flour was ready, she began to prepare the *ilende*, the vegetable loved so much by the Wagogo and which, together with *ugali*, represents the meal, that is, 'the food to satisfy hunger and to make the body grow'. At a certain point, speaking about women's agricultural tasks, she remembered when, as a young bride, she joined her husband's family. In her story, she presented me with a female image which overlapped the one I had already learnt to recognise from my first weeks in the village, and which was now emerging in a rather complex and articulated form.

Living in close contact with the women allowed me to observe them in the 'always the same' daily reality marked by the repetitive rhythms of domestic tasks: house cleaning, preparation of the single daily meal and milking of the cows. Other duties must be added to those mentioned: the long walks to collect wood every two or three days, and the daily or, according to the site of water sources, every-other-day walks to collect water. Furthermore, participation in the weekly markets of Cigwe or Dodoma, as sellers of saucepans, groundnuts or chickens, represented a change to earn a little money, by preparing *ujimbi*, the beer which, when not consumed on feast days or at harvest time, is sold in the village.

An atmospheric event – the rains at the end of November or the beginning of December – must be grafted onto this temporal view. It marks the start of the agricultural activities and gives a different setting to the women's tasks, taking them far away from home, even for whole days at a time in the first weeks dedicated to hoeing and sowing. At the end of these tasks, their presence in the fields becomes sporadic, until it becomes daily again at weeding and at harvest time.

In her role as wife and mother – and over and above her domestic, agricultural and milking activities, as well as small commercial enterprise – the woman is also responsible for the care and rearing of her offspring and the care and health of the family nucleus, in an interlacing where her productive and reproductive abilities repeatedly intersect. So, when a woman is engaged in agricultural work and is breast feeding her few-months-old baby, she involves him completely in her day. This occurs daily for most of the breast-feeding period, and above all when the child is not yet able to walk and his nutrition is completely based on mother's milk. In this period, the baby accompanies the mother, tied to her back, every moment of her day. When sitting, she holds the baby in her lap, by simply rotating the *kanga* holding him towards the front. Or she lies him down on her legs held straight out in front of her when seated, as is African custom.

These are the moments, as I often saw, in which the mother plays with her child, bouncing him or caressing him, tenderly repeating the little cries made by the child in answer to her gestures and trills.

In spite of the daily work load, the women do not hesitate to welcome visitors, other women mostly, with whom they interrupt the rhythm of work in favour of these relational moments they never seem to refuse. I learnt, in fact, that these moments were considered part of their daily rhythm.

This feminine world, which appeared to me from the very first months of my stay in the village, was gradually offering me, with its own peculiar features, a rich and variegated image of the women, both in relation to the network of family, relatives and neighbours, and to the roles a woman is asked to cover during the course of her existence. In fact, from birth and throughout the years, each individual of the female sex (in the same way as an individual of the male sex), gradually assumes different positions in the network of social relations: firstly, she is daughter, sister, half-sister, grandchild, niece; then she is wife, daughter-in-law, co-wife, mother, aunt; then last of all, mother-in-law and grandmother. In these different positions, she must interpret more than one role, she must assume more than one identity contemporaneously, each one having its own cultural, social and psychological expectations, each one having its own behavioural rules, in which she must be practised.

Therefore, to ask oneself what kind of woman – what kind of wife or mother – the Gogo female child will be, is the result of an educational process of which she has been an object since early infancy, and which reaches its crucial moment in the rites of puberty.[21]

Even though these rites do not have the emphasis of the past, as my elderly informants told me, the first menstruation is still an important moment in the life of an adolescent, as it marks the passage to social and physiological maturity, and therefore, to her *future* being as wife and mother.

Once upon a time, the women in this neighbourhood – [the elderly woman told me] – all ran to the sound of the drum and danced even for two nights and two days in honour of the girl – *wakutowa ng'oma yakalagala*,[22] they dance for the girl who gets her first menstruation – beer and food was abundant. It was a special way of dancing, for a special occasion – *wakuvina lihedule* [*kuvina*, to dance, *ihedule*, a special dance for a girl who gets her first menstruation] – as a sign of joy, *nyemo*.

This visible change in the physiology of a young girl is still an important event on the personal level, and as a feast, on the social level too, with the participation of the women in the neighbourhood. It is the moment in which the elderly women impart some teachings, *zingani*, to the young girl, about how to behave, particularly with the opposite sex, as it is necessary to avoid pregnancy before marriage, *kutumula*.

Particular attention is therefore shown to matters regarding the sexual sphere – traditionally a secret knowledge (*mizimu*) – including instructions on the techniques of sexual relations, on conception, on childbirth, on the behaviour to be kept with males and with the future husband. In the role of future mother, she is taught how to care for children and about the importance for the family of healthy children who will become healthy adults. The teachings are complete with rules, prohibitions, sanctions and the possible consequences should the traditional precepts, above all in the sexual sphere, be disregarded. In this way, the community's values and expectations of how a wife and a mother should be are transmitted to the girl, directly and indirectly.[23]

When a young woman gets married and joins her husband's family, she is aware of the rights and duties that her new status imposes on her, and when later she becomes a mother, she knows that she must concentrate on caring for the children, so that she can prove to her husband, to the other women of the homestead – her mother-in-law above all – to the other women of the neighbourhood and also to herself, that she knows how to give birth and to raise healthy children, in short, how to be a 'good mother'. This commitment is not, however, in daily life, an alternative to her role as a wife and to the duties that go with it. While her condition as a nursing mother is exalted, it does not, however, exempt her from working in the fields, from collecting water and firewood, from crushing millet, from cleaning the house and the courtyard, from milking the cows, from preparing food and from taking care of the other children. These are the tasks which are traditionally entrusted to women. Then why highlight and pay so much attention to her role as a mother?

First of all, because when a woman has given birth to a child, she has fulfilled her task of reproducing new beings for her husband's line of descent. Naturally, having brought a new individual into the world, motherhood is only the first act of the articulated process which is mothering, starting from *knowing how to give* 'good' nutrition. It is precisely with breast feeding that a woman starts a new period, marked by a series of rules, obligations and taboos, but also by expectations, and she will have to learn to make these cohabit with other precepts, rules, obligations and expectations. As we shall see, these, in turn, are also in conflict, one with the other.

Notes

1. The *kaya* (sing. and pl.) represents the domestic group, the composition and extent of which – from the elementary family to the polygenic family or extended patrilinear family – depends on its development cycle (Rigby, 1962b; 1969b).
2. Rigby (1966a; 1969b) speaks abundantly of the Wagogo houses and Beidelman (1961), on describing those of the Baraguyu, highlights the analogy with those of the Wagogo.
3. *Igagala*: inner ring of cattle byre (Rigby, 1969b: 171). My assistant told me that sometimes this term means an incomplete house: 'Maybe someone was building a house and did not complete it. This is *igagala*'.
4. I sometimes saw palisade-type fences, in which the entrance was marked by two large poles and a third on top.
5. *Yisi* indicates the 'ritual area' or a geographically delineated country. Rigby writes: 'These ritual areas are populated by homestead groups, which move fairly freely through them and whose members belong to a variety of clan affiliations' (Rigby, 1969b: 14).
6. The dual system of classification is rather common in sub-Saharan Africa and in particular among the Bantu-speaking populations, as many studies have testified, beginning with the classics, Evans-Pritchard (1953; 1956), Needham (1960; 1973), Beidelman (1961).
7. To counter possible threats from the outside, special medicines are placed in the centre of the area and at the four cardinal points, zones which are recognised as the ideal territorial bounderies, to form an invisible 'fence', an impassable protective barrier (Rigby, 1966a: 6).
8. When a family group moves, it reuses, as far as possible, the wooden structure of the house which is being left, in particular, the strong beams supporting the roof, *isumbili*.
9. Mlindimo and Iweta are the two hills which can be seen at a few hundred metres from the dirt road, near the village coming from Dodoma. Both of them emerge in isolation on a wavy terrace, the first to the left of the road and the second to the right. An authoritative elder indicated them as special places when telling me the history of the village.
   Mlindimo is a hill covered with intricate vegetation, passing from burnt brown to intense green with the first rains. One day I was accompanied towards the summit where a large and dark cave, hidden from indiscreet eyes, was described to me as the home of the spirits of the dead. The Wagogo practise rites dedicated to them, just outside this sacred place.
   For the inhabitants of Cigongwe, Iweta represents continuous proof of the anger of the ancestors. It is said, in fact, that long ago, annoyed by the neglect shown to them, they demonstrated their anger by breaking up the mountain into pieces, as a continuous reminder of men's indifference. Its isolated profile is very different from all the others and really gives the impression of something disturbing having altered its shape. Five large white masses, shaped like oblique prisms and parallelepipeds of different heights, rise towards the sky from an embankment which is green only in parts. They rise from an indescribable disorder and overlapping of rock pieces and splinters, as if after an explosion. It resembles the grimace of a mouth without teeth.
10. In his interest in the ways of using space and the meaning attributed to them, Hall shows how the distance which individuals impose between each other,

in different moments of encounter, assume different values in different cultural settings, just as in the use of space (Hall, 1966).

11. The frame of the dwelling has a specific nomenclature: *masumbili* (sing. *isumbili*) are the internal and external vertical supporting beams; *mchichi* are the horizontal beams which are placed along the top of the walls, designing the perimeter of the hut; the beams and the planks of the ceiling are placed on these; *mahapa*, medium-strength planks which form the roof, between one beam and another; *walo*, small planks which form the frame of the roof and on which hooked dowels, *ndobogo*, on closing the frame, help to maintain, during the violent downpours, the plaster, *ilongo*, which covers in various layers the lattice composing the roof. As well as the supporting beams, the walls are composed of a thick series of small poles, *izengo*, which are placed in parallel, at a short distance one from the other. This vertical series of small poles is crossed horizontally, at about midway along their height, by a pair of plaited shrubs, *sito*.

12. I only once saw the walls of the *ikumbo* embellished with some geometrical drawings, running along the wall just under the roof, as a sort of cornice, and below them some animal reproductions. Although simple in design, the giraffes, elephants and gazelles reproduced gave the impression of wide spaces and an elegant gait. In one other case, I saw large letters on a wall, tracing a phrase of welcome.

13. Even some utensils and instruments of domestic work are assimilated to male and female. For example, the mortar, a container, and the pestle, with a vaguely phallic form, are respectively female and male, in the same way as the sticks for the fire. Those on the ground are female, concave in shape to collect the slivers rubbed off the male one, held between the hands in a vertical position (Rigby, 1966a: 3–4).

14. This piece of black cloth is still used for ceremonies requiring the intervention of the spirits of the dead. Black, *utitu*, is a propitiatory colour, as it is for other pastoral populations of East Africa. It is the colour of the clouds and the sky which announces rain, beneficial to men and herds. I myself wore the black cloth when assisting in the sacrifice of a black sheep, in front of the cave of the spirits of the dead on mount Mlindimo. Rigby (1966a) remembers other ritual moments in which black is the propitiatory colour.

15. *Uhemba*: in Dodoma area and specially in Cigongwe is *uwele*.

16. Among the vast literature on the theme, Bourdieu's study on the *kabyle* (1990) is a classic; the work of Carsten and Hugh-Jones (1995) is an interesting example of the rereading of a rich and stimulating subject.

17. Because of the importance of the *nyumba* as an economically independent unit in the development cycle of domestic groups, Rigby defined it the 'matricentral unit' (Rigby, 1969a: 169). With this definition, he intended to highlight both the pulsing heart of the domestic group, and the ambit within which the conditions of the successive divisions of the family units develop, in the incessant process of development and fission of the *kaya*. The article by Smith Oboler (1994) is interesting in this regard.

18. Among the activities undertaken by males remaining in the village, the most remunerative is the sale of wood charcoal. In the dry season, it is possible to see sacks of it along the dirt road leading to the village, waiting for a pick-up to take them to Dodoma. Another traditionally male activity is honey gathering. I also saw men engaged in artisan work, constructing the large granary for *ikumbo*, or carving chairlegs to be taken to Dodoma, and a blacksmith shaping what I took to be a blade, from a matrix made from a goat's stomach.

19. *Itembe*: in Gogo terminology, it indicates the wings formed by the *kaya*, giving it its typical horseshoe or L-shape. Rigby affirms that there is a *nyumba* in each wing, but I was able to observe, in the most extensive living units, the presence of more than one *nyumba* for each wing of the building. In any case, this specification relative to the term *itembe* goes in the direction followed by Rigby (1969b, note 4: 157), when he called attention to the erroneous extension of the term to all *kaya* by other authors, for example Cluass (1911).

20. *Bibi* (pl. *ma-*) is also a grandmother.

21. The rites of puberty mark the moment of passage to adult age and represent what are for the males, initiation rites. At around the age of 9–10 years, female children are subjected to initiation rites which include clitoridectomy, but which do not have much social resonance.

22. *Yakalagala* is the term used to indicate the condition of a young girl only at her first menstruation, from the verb *kukalagala*, to get menstruation for a young girl for the first time. Immediately afterwards she is *mnyacipale* or *mwanyacicane*. The menstruation, '*hono yatema ihamha*', literally, when she has cut the leaf, is the expression used by women to indicate this physiological event, to which social recognition is given, in honour of menarche, equal to circumcision for males.

23. From the classic work by Richards (1956), Turner (1967), and La Fontaine (1972) to the more recent works by Ahlberg (1991; 1994), we can read about the force of tradition in the setting of rules aimed at sanctioning correct sexual behaviour.

CHAPTER 3

BREAST FEEDING

Breast feeding between nature and culture

One day, when I was visiting a homestead, chatting with men and women about the harvest of peanuts being picked from the plants, a small boy of about three years arrived and settling himself on his mother's lap, took her breast in his hands and began to suck. The woman smiled at me and, in no way surprised or annoyed, serenely continued her work. After a few moments, the child got up and ran off to join his companions. What I had just seen was in no way unusual; I knew about the long breast-feeding period, on average between 24 and 30 months, and I had repeatedly observed the mothers' willingness to allow their young children access to the breast. Furthermore, I knew that the quality of this dependence between a mother and her child allowed them to experience, for a few months, a symbiotic relationship, foreign to the behavioural models of industrialised societies. However, on observing this scene, I could not help going back in my mind to my first weeks in the field, when I began conversations with the women by asking them if they breast-fed their babies. They looked at me with ill-concealed amazement, even bewilderment, as this strange and senseless question had for them such an obvious answer:

> How should I feed him? My mother, my mother's mother, my sisters, my neighbours, all of them have breast-fed and breast-feed their children. It is the woman's task to give their milk to their children. Otherwise why should there be milk in women's breasts? It is natural for a mother to breast-feed her children.

These considerations revealed the meaning of 'giving', of 'offering' the breast to their babies, from the very first hours of life, permeated by the inevitability of physiological processes and, at the same time, by the equally strong ineluctability of the processes of inculturation which, once internalised, dictate how individuals see the world, behave, feel and perceive emotionally.[1] Therefore, these affirmations made me think about the infancy of my interlocutors, when, as children, they observed the life around them and learnt to become part of it.

Early on, they accompany their mothers or elder sisters to fetch water and to collect firewood, they help in keeping the courtyard clean, they carry the youngest child tied to their backs and they see their mothers and the other women of the *kaya* looking after their children, as they pass from one pregnancy to another, with long intervals of breast feeding. In this way, as they grow up, the idea of having their own children to look after and to breast-feed is, for young girls, an aspect which is closely connected to their future as adults :

> When I was a child, and above all, when I became *mhinza*, a marriageable girl, I looked with envy at the older girls, married and already with a baby to feed and grow. The desire to have a child of my own to feed and grow, became stronger and stronger in time.

This kind of affirmation was repeated when I attempted, talking with the mothers about their daily tasks, to compose a framework, within which the concept of breast feeding developed and took on meaning. On these occasions, I tried to understand which tasks the community assigned to them, what their growing expectations were and what elements made them feel fulfilled. In this way, I was able to focus in marriage and maternity, both of which inevitably followed puberty rites, the two steps which allowed them by right to become part of community life in every aspect.

On speaking of maternity, they all emphasised, more or less explicitly, just how much breast feeding responded to a custom, which linked the physiology of one's own body with a child's nutritional needs. Their words implied the strength of the biological bond between a woman and her offspring, a bond which begins with conception, continues through pregnancy and which is prolonged after birth through breast feeding. The conviction of the *naturalness* of the presence of milk in a mother's breast was so rooted in them that if by any chance it was not present after birth, this would unanimously be declared the result of the external intervention of witchcraft, dictated by the jealousy or malevolence of other women, or of a rejected lover.

I found, therefore, in the unanimous response of the Gogo mothers, the same widespread Western stereotype, which puts many aspects of female physiology – menstruation, birth, menopause – under a halo of necessity, because they are determined by laws of nature and, therefore, are strongly instrumental in defining the roles and tasks of women within society.[2] And yet, on continuing my work and on hitting on the right questions, that is, those which *made sense* to the women, on learning to listen to what they were really saying, breast feeding took the guise of a 'social phenomenon', in the meaning evidenced by Marcel Mauss in his essay on gift (1965: 286) – a 'social phenomenon' able to communicate multiple cultural meanings, multiple ways of feeling and perceiving, on which fundamental rules for the social organisation of communities and their institutions act.[3]

Starting from the Gogo women's conception of breast feeding, as a *'natural' maternal response to the baby's nutritional needs*, my task became, therefore, to understand the presence of those cultural and social aspects which transform a physiological potentiality into a social fact. I intended to grasp the *sense* and the *value* of a process, at the same time *composite* and *complex*, in breast feeding.[4] *Composite*, because it is the product of a combination of different elements, that is, of the interaction of both *physiological* elements – health, nutrition, maternal milk, let-down reflex – and *cultural* elements – beliefs, aptitudes, customs, behaviours and expectations. This interweaving of the biological and the cultural makes the bond between the mother and her baby, understood as a nutritional act, only one of the aspects of breast feeding. In the absence of pathologies, it depends on motivation and knowledge, on opportunity and contingency, experienced and experimented by the mother. Yet again, it depends on psychological mechanisms linked to the sphere of the emotions which unite the mother and her baby, and the latter to the source of food, satisfaction, warmth, protection and stimulation (Quandt 1995; DeLoache and Gottlieb 2000).

> I like it when my child seeks comfort in my arms.
>
> I like it when he holds on to my breast seeking protection. It is a moment in which I don't feel myself as just a source of food.
>
> I like to feel him so close, holding on to me so tightly.

With these words, some Gogo mothers demonstrate how the nutritional aspect, which is linked to an imperative imposed by the 'nature of one's body', cannot be separated from the emotional aspect which is nourished by affection, love, tenderness and sensitivity towards the needs, not only nutritional, of one's own offspring. These ways of feeling, of relating to one's own child, at the same, are part of being woman and mother, where individual personality and roles interweave along cultural lines, sometimes fragile, sometimes rigorous. The words of Patricia Stuart-Macadam become, therefore, more than ever appropriate in defining breast feeding as *'the ultimate biocultural phenomenon; … not only a biological process but also a culturally determined behaviour'* (Stuart-Macadam 1995: 7).

The interrelation of the two levels, physiological and cultural, must, therefore, conjugate with a further complexity. In fact, breast feeding is also a *complex* process, because the link between the physiological and the cultural participates in different, but closely related, levels of social organisation, for example, sexual roles and *gender*, social relations, kinship systems and division of labour. It follows that, although the daily bond between mother and child seems to assign her every decision, breast feeding is not managed *in proprio* by women. Furthermore, it is not promoted *tout court* by the needs of the infant; there are other responses to his cries. Both are affected by the action of the ideas, behaviour and expectations that a given society has placed, not only on the female body and on the development needs of the infant, but also on the rules and prohibitions, aimed at defining the different roles that women are called upon to perform and, through which a community understands the organisation of its own associative life.[5]

Conscious of this complex network of relations and interrelations, the Gogo women's expected answer to my question, (in no way provocative), whether they breast-fed their babies, opened the way to a study of the intricate link between nature and culture, once again one facing the other, or better still, the one correlated to the other (Goodman and Leatherman, 1998).

The modalities of breast feeding

Exclusive breast feeding

The newborn are breast-fed shortly after birth, although the time, in terms of a few hours, may vary according to the situation. The majority of women give birth at home, helped by other women of the family or neighbourhood, and also by the mother, if she does not live too far from the village. The custom of giving birth to the first child in the home of one's family of origin, is still respected, if not generalised. The intervention of a traditional birth attendant, *mung'hunga*, is decided by the elderly women of the *kaya*, in first place by the mother-in-law, and is solicited, in any case, if labour is long or painful, or if there are complications.[6]

Immediately after birth, the mother and baby stay beside each other, resting or sleeping. While the mother gets back her strength, the baby may begin to cry, warning the mother of his need to be fed. The child's cry is unanimously understood by the women as a sign of hunger, hunger to be satisfied by maternal milk: the food of the newborn. In these first relations with the newborn, the mothers are aware of a circumstance which is rather neglected in the West, since breast feeding has been regulated by tables and performance times: breast feeding is closely connected to the sucking of the newborn.[7] 'By sucking, the baby "calls" the milk', was what the women repeated to me, a way of expressing the fact that the action of sucking and the onset of lactation are understood to be closely correlated.

This relationship, fruit of observation and personal experience, handed down by generations of women, and not only Gogo women (Palmer 1990; Maher 1992; Dixon Whitaker 1994, 2000), was well present in the mothers' minds, so much so that, if the baby did not take the nipple shortly after birth, then the mother herself would stimulate it, delicately rubbing it to 'call' the milk for when the baby would want it. When describing the manual stimulation of the nipple, they use the verb *kunyaga*, which describes literally the rubbing of clothes in water when washing them.

If the baby is quick, however, and insistent in demanding milk and this is 'not yet present' in the breast, that is, it does not arrive with the urgency required to meet the demand, a helper will prepare boiled water with sugar or salt. It is worth noting that the use of sugar or salt is rather recent and, in any case, not generalised. The use of salt is probably an extension of information received in the clinic for the treatment of diarrhoea in a child.

Before starting breast feeding, the newborn may also receive a mouthful of soup of millet and water, to which medicines may possibly have been added to help the mother recuperate the energy lost during birth. The offer of this food to the child is totally symbolic. It is a sign of welcome to the family, a first gesture of recognition as a member of the community.[8]

I was very rarely able to assist at the slaughtering of a goat, as is the tradition when a child is born, and from which a nourishing meat broth is made, *muhuzi*, and given to the woman to help her regain her strength after delivery (Rigby, 1968b: 159). When I did happen to see this event, I never saw the animal skin being used to make the *sambo*, the sack which was used in the past for carrying the baby on the mother's back. Today, all the women use a *kanga*.

One day, however, I did get an idea of this means of transport, on talking with an elderly woman about her first pregnancy and delivery. During the conversation, she suddenly stood up and fetched the skin of an ox on which pieces of pumpkin had been laid out to dry in the sun. She began to shape the *sambo* by using a corner of the skin, giving me an approximate demonstration of an object which I realised today's mothers had never received and probably never seen. It was basically a rectangular sack, sewn together, the woman informed me, with strips of skin, with two lower corners left open to allow the child's legs to pass through and rest on the mother's hips – in other words, something not very different from the modern baby carriers used in the West to carry children on the back.

I was often told that these traditions had been abandoned mainly due to the increasing lack of animals, and to the economic impossibility of buying a goat.[9] In fact, even if they do have goats, they do not necessarily sacrifice them, even for occasions like a birth, which, in other aspects, is marked by traditional practices. A practice still in use, for example, is that the father of the child should go to a diviner to obtain amulets and special remedies to protect the health of the newborn. I often saw newborn babies, a tiny leather sack around their necks, containing special medicines to keep away illness and evil, or with a sort of belt made from beads or shells tied around their waists.

As babies are breast-fed within few hours of birth, all of them receive colostrum, *mang'handa* or *mele ga mwamlwizo* or *mele ge sonje*, the 'first milk'. The mothers do not know about the important nutritive and protective qualities of this milk, and the consequent extraordinary benefits for the baby's immune system and, therefore, for the whole process of growth of the newborn.[10]

> The baby must suck the first milk, otherwise the *real* milk will not come out. The *real* milk is white, *mele mazelu*, while when it starts to come out for the first time after delivery, it is yellowish and dense like *manhandu*,[11] the cream of the milk after boiling.

This was the most frequent reason given me by the women to explain the consumption of the 'first milk', of a colour and consistency so different from *real* milk, the one used to feed and help the baby grow. Others added that every baby has its own colostrum, which it feeds on while in the mother's womb. I did not take this statement any further at the time. I remember the women who had spoken about this becoming silent when I tried to understand better what they wanted to say. It was during the first months of my research, and I felt that I should not force my interlocutors, as I may have inadvertently touched on aspects of their behaviour and beliefs which required caution and a specific strategy in dealing with them. The above-mentioned statement remained, in that moment therefore, just a note in my jotter. It came up again, a few months later, and in a completely different context.

I must add here that very few women diverged from the above statement about colostrum. Only one woman told me that she had thrown it away because it was 'bad milk'; a few others told me that they had sucked it themselves because their breasts were full of milk before the baby showed signs of wanting to suck, while a few others associated it with the evacuation of black faeces (*yakuhalisha mabi ne mapuko gakwe*), meconium, an effect recognised by medical science (Mohrbacher and Stock, 1997: 22).[12]

A long period of close and continuous life together for the mother and the newborn begins with the start of breast feeding. The baby accompanies the mother wherever she goes during the day, and the mother satisfies 'on demand' the baby's need for milk. During the night, the mother and newborn sleep together, sharing the modest pallet in the *kugati*. The mother tends to sleep on her side, thus facilitating breast feeding. The women told me that, as the months go by, this position allows the child an autonomous access to the breast, and he feeds himself during the night.[13]

The complete willingness of the mother to offer her breast to the child made it very difficult, if not impossible, for me to determine the number of daily feeds. 'The baby doesn't feed only when sleeping', was the recurrent answer to my question about the number of feeds per day. Entrusting myself to daily observation, I noted that, in general, the newborn frequently sucked for three, four or five minutes, although I also noted it lasting more. The women compare the frequency with which babies seek the mother's breast to the continuous oscillation of a swing. In fact, to explain this, the women use the verb *kitundya* (literally, the backwards and forwards motion of a swing). They further extend this meaning, to the action of cradling, a calming gesture, which is often accompanied by offering the breast.

My interlocutors used many different expressions to express the concept of breast feeding 'on demand': *mbilimbiliti, cibitilita, mapango gogose*, literally 'every moment', 'continuously', 'every time that', or *likuswa, liwile lizuwa, kutwa du*, respectively, 'the whole day', 'until sunset', 'from morning to evening'.

In the nutritional model of breast feeding, which the women have made their own through processes of inculturation, where observation and imitation have a significant role, the meaning of crying as a request for maternal milk, is, as has been said, shared by all. This conviction is so deeply rooted that, on many occasions, the women emphatically underlined the ability of the child 'to call out' their milk by crying, even in the absence of the mother. In this way, if a woman finds herself for some reason at a distance from her little one, and sees her milk is spontaneously leaking from her breasts, she must return home immediately, as that free flow of milk signals that the baby needs to be fed.[14]

We must not neglect, however, the mother's awareness of her infant's desire for the breast being motivated by the need for comfort, tenderness and security. She considers these needs as just as important as the nutritional ones and, therefore, her promptness and willingness to offer the breast are no different. On more than one occasion, I saw the breast being offered over and above the need for food. For example, when a child fell on taking its first steps, or hurt itself playing, when it was tired or sick, or afraid in the presence of a stranger, as often happened during my first visits, all of these were good reasons for seeking the maternal breast. It was therefore not unusual for me to see a child seek refuge in its mother's arms, while she was sitting intent on some task or other, or while chatting with other women, and suck for a few minutes, or just keep the nipple in its mouth or bury its head in its mother's breast, in search of shelter and protection, or simply to satisfy a pure need for tenderness.

In the first months of the child's life, the mother's promptness and willingness to offer the breast in response to the child's cry is important for two different and antithetic aspects. The first is positive: a strong bond between the mother and child is created, because the child is fed when it needs it and because, by offering the breast on demand, she responds to the child's cry for comfort, reassurance, help and affection. The second is negative: since she responds above all to the child's cry, the mother may delay feeding when it doesn't cry, when it is apathetic or without appetite, perhaps the first signs of an oncoming illness. A Gogo mother is generally not, in fact, ready to stimulate the child to feed, as much as she is to respond to his crying. In this way, a vicious circle may be created, in which the reduced desire to suck makes the infant more apathetic and therefore less and less desirous to seek the breast. This behaviour, by reducing the let-down reflex, threatens the whole breast-feeding process, from the health of the infant to the production of milk (Barness, 1993b; Stuart-Macadam, 1995; Campus et al., 1998).

The various activities that the mother must see to during the day, and above all when she carries the lastborn with her during the period of agricultural work, tend to be moments in which the time between one feed and another is stretched, if the child's demand is lacking. To believe, therefore, that the child is the best judge of its own nutritional needs may mean running risks, even in a situation in which the mother is never annoyed or bored by the continuous demands of her child, for the long period of breast feeding.

On concluding these observations on the modalities of breast feeding, I remember that the women make no distinction on a sexual basis in their behaviour with infants. 'The needs and desires of the newborn are the same, without distinction between the sexes. Males or females, they must both, first of all, grow', all of the women told me, extending this full parity, as we shall see shortly, to weaning practices.

From exclusive breast feeding to mixed feeding

The mothers, although fully aware of the essential role of their milk, are just as convinced that it is not sufficient for the nutritional needs of the infant for the entire period allocated to breast feeding. As the months go by, the infant's physical development will require increased quantities of food, for which their milk alone is not sufficient. *'Mele si gakukama'*, the women say, intending with this expression to indicate the inability of their milk to satisfy the nutritive needs of the child as it grows.

The verb *kukama* literally means 'to milk', but it is also used to indicate the action of plastering the walls of the huts with clay. It is in the latter sense that the women use it in the above-mentioned phrase: the milk is no longer sufficient to plaster the walls of the child's stomach, it does not adhere to the walls, thereby failing in its function of satisfying the child's hunger. The child has grown, is bigger, *yamemamema*, and with him, his stomach has become more exacting and strong enough to receive semi-solid food, *itumbu lyakangala, ng'hkwanza mpela nhili*. I repeatedly heard, when discussing the modalities of breast feeding with mothers, first of all, how important and irreplaceable mother's milk is for the child's life, and then 'but not for long'.[15]

Gogo mothers, however, perceive this need, as we shall see, in advance of the times fixed by medical science. They told me: 'When the infant is *yakali mdodo*, mother's milk is the only food, but when he is *yalasuga*, he needs *nhili*. Later, when he is *yalawahapa* he eats *ugali na ilende*, the food of the Wagogo'.[16] If these are the different periods to which the women theoretically link the passage of age of their children, in terms of nutritional needs, they made it clear that, on one hand, the various dietary periods were determined by the infant and, on the other, that in these three different phases of growth, maternal milk remained the privileged food, essential for his health and well-being.

A mother is warned of her child's dissatisfaction with her milk by a more and more insistent cry and by increasing fretfulness between one feed and another. 'My child is starting to give me problems, *akunyuzu*', one woman confides in her female friends. This 'giving problems' refers to a nervous baby, frequently subject to bouts of crying, both of which signal the hungry child, the child who needs more food.[17]

The need to prepare other food is yet another thing for a mother to add to the list of tasks she carries out during her day. The introduction of supplementary food begins, according to my interlocutors, around three or four months, and is initially offered to the infant on a sporadic basis. It becomes more continuous and systematic over the following months. Normally, it is only when the child is able to sit up on its own, *honoyonza kikala hasi'* (between eight and nine months) that supplementary food is given every day and more than once a day.

Although most mothers normally begin to give food other than milk around the age of four months, I did encounter exceptions, some at less and some at more than four months. Some women were proud of the fact that they breast fed exclusively up to nine months or more, thanks to their 'good and abundant' milk, while others, a small minority, told me that they had given solid food at one or two months. One mother confided that she had begun to give solid food once a day when her child was only one month old, just as a precautionary measure. She wanted to have an alternative, if, for any reason – illness, refusal of the breast, or anything else – her child did not want to, or could not, suck her milk. This behaviour was linked more to the desire, which most mothers shared, to familiarise the child, from its very first months, with what would be its principle source of nourishment once weaned, *ugali*, than to the insufficient nutritional value of her milk.

Amanya kulya ugali, 'learns to eat *ugali*', or *akusacila ugali*, 'creates the desire, the need for *ugali*', are expressions with which the women underline the necessity to make the child 'learn' the taste of food, to 'get him used' to its flavour, to 'bestow' on him the desire for food, for that specific food to which the Wagogo entrust their survival. This was directly confirmed when I saw mothers, during their meals, feed their child of about one year with small mouthfuls of *ugali*, only *ugali*, *ugali wacilumanga*, that is, without vegetables.

The first food is a rather diluted soup of millet flour and water, *ubaga cinyetumba*, which the baby starts to suck from its mother's fingers, or from the palm of her hand which forms a spoon shape. This food, colloquially called *uji* in Swahili, initially eaten once a day and not every day, does not significantly alter either the baby's habits, or those of the mother, even though for the baby's digestive apparatus it represents the first break in the optimum equilibrium offered to its organism by maternal milk.

With the passage of time, and compatible with the child's nutritional needs, communicated to the mother through its cries, *uji* is given every day, at the beginning once a day, later more than once. The mother cooks it in a special utensil, *cipeyo cinyan'hwi*, early in the morning, where it remains conserved for the rest of the day.[18]

Now, *uji* is denser and can be enriched with other ingredients, according to what is available. The most common among these is the flour obtained from finely ground peanuts, or *nzugu*, (bambarra groundnut), products cultivated by the women both for domestic use and to sell in the market when necessary. Another important ingredient, rich in protein and easily obtained cheaply, is *dagaa*, dried sardines ground into flour. According to the seasons, mothers may also add wild fruits and, among these, *upela*, the fruit of the baobab tree, is the most common. They rarely substitute water with cow's milk. In the past, when the majority of the Gogo families kept cattle, the women frequently cooked millet flour mixed with fresh milk, *mele masusu*, or with soured milk, *mele masuce*. As an alternative to milk, and when available, they added butter, *tilazi*, obtained by churning the milk in a pumpkin, *nh'oma*.[19] If the mother does not have time to cook food for her child, she will use *upwa*. This is the name given to the mixture of water and flour which the woman quickly extracts, using a half-pumpkin gourd, *lupeyo*, from the *ugali* which she is cooking, just as it begins to boil.

The start of weaning and the weaning process itself are critical moments in the life of a Gogo child. Although the women say that as the months go by, they add other ingredients to the mixture of millet flour and water, and that the child receives more than one meal a day, in reality, and as I was able to observe, these intentions are not constantly put into practice, and sometimes the child is not adequately nourished. It is true that the difficult habitat means that resources are limited and it is probably because of this that the older women scold the young mothers for not preparing the traditional *ubaga we zaliko* or *ubaga iloweko*, a very special food for the good growth of children. The peculiarity of this food is due to the fact that it is composed of flour obtained from germinated millet.[20] This special food takes time and effort to make, both of which, according to the older women, are lacking in the young mothers. The process takes some days: first the millet seeds are soaked in water to germinate them, then thoroughly dried and ground to give a fine flour. The older women told me: 'It is so rarely prepared that the child does not know the taste and refuses it or eats it unwillingly, consequently the mothers are discouraged and do not make it.' These considerations on the taste are understandable when we consider that when the women prepare the flour, they cook a quantity for a few days. Being left for a few days in a saucepan near the fireplace means that it is subject to fermentation and assumes a sour taste.

During the second year of life, the child is fed *uji* two or three times a day on average. If the child eats without problems, the mother may consider leaving him at home with his older brothers and sisters when she is busy in the fields or needs to go to the market. The older children become babysitters, *mkoci*, looking after the younger brother and feeding him the food prepared by the mother in the early morning. In this situation, the food reserved for the smallest child is often shared by the others, thereby significantly reducing the amount for the most dependent member of the group.

An ever-increasing consumption of additional food, together with the absence of the breast for many hours, causes a progressive and inevitable decline in lactation. The reduced frequency and avidity of the child's sucking inevitably affects the dynamics of the let-down reflex. The child continues, however, to seek the breast, for nutrition and comfort. The mother does not change her attitude towards the child, continuing breast feeding on request, and is available night and day when present. This situation continues until the breast is taken away completely, a totally new phase for the child.

In fact, although the first phase of weaning has all the characteristics of a process through which the child begins to consume solid food from time to time, until it progressively becomes part of its diet while still breast feeding, in the second phase a mother suddenly decides to end breast feeding.[21] The verbs *kwima* and *kuleka* used in these circumstances under-line the act of stopping, of finally terminating breast feeding.

The end of breast feeding

The removal of the breast is a decision that the mother makes autonomously. There is no established time, *hasina kilinda*, to stop breast feeding, although it is preferable for it to happen when the child is between 24 and 30 months old. The mothers behave in the same way with both male and female children. When they feel that the child has reached 'the right age', or 'is old enough', or 'is able to walk quickly', that is 'the right moment'.[22] The women do, however, allude to themselves, saying: 'When I feel tired, *nakatala*, that is the time to bind my breasts, *ng'huwopa titombo.*'

The attitude may be different towards the lastborn, the probable last child, as an elderly informant told me. She said that she had breast fed her last child until he was almost five years old, for her own pleasure and that of the child: N*ani ane aa, cawa ca udala, nomosa du mihano nhondo.* On the other hand, over and above possible exceptions and motivation dictated by a certain idea of the child's development, or by his autonomy of move-ment, the desire for a new pregnancy plays, without doubt, an important role in a woman's decision to wean the lastborn.[23]

When a woman decides to stop breast feeding, she does not go back on her decision. It is an unpleasant surprise for her child: suddenly he no longer has access to the breast, neither for nutrition nor for comfort. The most important problem for the mother to solve is therefore to 'make the child forget the breast', 'to stop the desire for the breast', so that the child no longer cries or seeks it.

In order to do this, they may go to a traditional healer, male or female, who in this case is called *msilici.* The expert knows the medicines, *miti*, lit-erally trees, which are able to make the child 'forget the breast'. The healer himself generally cooks these roots with millet or sorghum flour, or he gives the fine flour of the roots to the woman to add to the child's food.

The women know various other methods to discourage the child from seeking the breast, however. One of these is a 'special' *ugali*, infallible according to them. I assisted in the preparation of this 'antidote' one morning early, in a *nyumba* in which I had been a guest since the day before. The previous evening, in a conversation around the fire, the woman had confided in me her desire for a new pregnancy. Her lastborn was then about three years old. We were in the *kugati* and I saw her take a small quantity, less than a mouthful, of *ugali* from the saucepan placed on the wall near the fireplace. Looking at me she said: '*ugali wono ugonile*' (*ugali* that had been sleeping), in other words, porridge which had been cooked the previous day. She then rose and reached towards the beams of the roof, taking down what seemed, at first sight, to be a piece of mater-ial stuffed between a supporting pole and the thick frame of sticks and red earth of the roof. When she came nearer, I saw that she was holding a packet made from the bract of a maize cob. She opened it and I saw small dark pieces of what looked like pepper grains, maybe small stones, I thought. As usual, there was little light in the room and I was merely guessing. She informed me that they had once been insects, more precisely flies, *hazi*. She took one of the grains and mixed it, or rather hid it, in the porridge, together with some wood charcoal, *mteme*, which she crumbled with her fingers. She began to paste it all together with her right hand for a few minutes until she had formed a small ball, in the centre of which she delicately shaped a small shell with her thumb. To my surprise, she poured a few drops of her milk into it, squeezing her breast with a pressing movement from the outside towards the areola. When she had obtained the quantity deemed sufficient, she approached the bed where the child was sleeping. She sat down beside him and placed the mouthful of porridge near her. She took the baby into her arms and softly waking him she moved her nipple to his mouth. Feeling the nipple tickling his mouth, he began somewhat unwillingly to suck. After a few moments, the woman rapidly and suddenly removed the breast and placed the 'enriched' mouthful of porridge into the baby's mouth, still open with surprise. With her fingers, she expertly forced the child not to spit it out, but to swallow it, which he reluctantly did. In the end, he burst into tears. It was done and with this 'mouthful' the child would refuse his mother's milk, or at least, this was what she expected.

Other women tried to persuade their children not to want breast milk by telling them horrific stories. Some, in fact, told me that they had told their children that they could no longer breast feed because tiny animals had defecated on their breasts while they were sleeping.[24] To confirm the truth of their stories, and in case the child should seek the breast anyway, they secretly rubbed pepper onto the nipple. In order to frighten them, two women told their children that a dog had 'sat down' on their breast during the night. This tale seemed to cause such disgust in the children that they no longer wanted to take the nipple into their mouths.

The majority of the mothers interviewed, however, told me that they rubbed their nipples with bitter substances which gave the child an unpleasant sensation when sucking, thereby causing rejection of the nipple. 'It's necessary to make the child spit in disgust, *kufunya*', the women repeated. The choice of leaves, fruit or roots, to be ground or cut up to make a paste, or from which to extract a liquid to be rubbed onto the breast, depends on the season. The women themselves look for remedies in the bush, with a special aim in mind: *Nimsilika yalece kulila* (literally, 'I'm going to look for (medicines) to stop the crying'). As a last resort, the woman may decide to send the child to the grandmother, generally the maternal one, so that he can forget his mother's milk, *mpaka kumgula mele gakuta*.

The diversity of these remedies confirms just how much the moment of 'binding the breast' is problematic for both the child and the mother. The latter knows, through direct experience or having heard it from other women, that this sudden passage to an adult diet, based essentially on *ugali* and *ilende*, will be difficult for the child to accept. It is for this reason that the women choose to wean the children generally during the season called *itika*, approximately at the beginning of March to the end of April, when, thanks to the rains, there is a wider variety of food than in any other period of the year. In fact, many types of pumpkins and courgettes mature in these months, *majenje, mahikwi, mayungu*, and are good complementary foods to the monotonous adult lunch. The women mash them and, if it is available, add sugar. *Ukoko*, scrapings from the *ugali* left over from the previous day, from time to time an orange or a piece of papaya, but above all ripe wild fruits represent a snack while waiting for a meal. A special food for the child's nutrition in this passage to adult food, not very common today, is *ugali* mixed with acidified cow's milk, *mele mphopota*.

The mothers claim that their children eat a lot when they are weaned. In reality, my daily observation showed this to mean that they eat a wider variety of foods when they no longer breast feed, and this in only a short period of the year. Once this period is over, *uji* remains the most common alternative to porridge, which quickly becomes the child's 'food' too for satisfying hunger.

The weaned child is given his ration of porridge in a different container from that of the adults. The latter take the *ugali* from the cooking pot, *nyungu*, with their fingers and, helping themselves with the porridge itself, they take the vegetables from a smaller pot, *cikubujilo*. The child, still too small to pick up hot food, eats his ration from a separate pot. By using his 'own' container, the child is facilitated in eating and the mother is able to gauge the amount of food eaten. It is no longer the child's crying which attracts the mother's attention to his nutritional needs, but the quantity of food consumed.

The women follow a very simple rule: 'If the child eats and leaves food it means that the portion was sufficient and the child is satisfied. If, however, he eats everything, it's necessary to give him some more to satisfy his hunger.' The simplicity of this rule suggests that the women do not sufficiently consider the child's eating ability and the fact that the porridge makes him feel full after just a few mouthfuls – both of which are causes of inadequate nutrition.

Weaning intended as the end of breast feeding gives rise amongst the adult women (and above all amongst the oldest, *bibi*), to severe criticism towards the new generations. They claim that they 'bind the breast' too early, when the child is still too small, *yakali mdung'hu*, literally 'it's still red'. This expression refers to a baby which is so small that it has not yet assumed the normal dark colouring. At the same time, they condemn the increasing tendency to entrust the child to a babysitter, *mkoci*, 'while he still needs his mother', meaning, he still needs the mother's breast.

The severest reprimand that the elderly women hold against the young mothers refers to the fact that they do not observe post-partum taboos, which can lead to the precocious interruption of breast feeding, *kulesa*, literally, 'to stop breast feeding'. When the women began to vaguely mention this, looking at each other as if unsure whether or not to speak about it, I understood that I should not take up the issue directly, but come to it transversely, only apparently moving away from the objective.

It is interesting to remember that when the women spoke to me about the general aspects of breast feeding, they always began from what they considered the normal, usual (natural?) *iter* of breast feeding. As they continued however, they always added 'although it is not always so', and from this statement I was able to dig deeper and ascertain that the global picture was richer, more articulated and complex.

I began to investigate, trying to understand the women's ideas about their milk, considered a fundamental nutrient, but at the same time, and early on, insufficient 'on its own' to satisfy the nutritional needs of their children. Through their ideas about the different characteristics of maternal milk and the changes which may take place in it, I was able to get nearer to, and then to face, the fundamental problems linked to maternal behaviour during the period of breast feeding.

Breast milk

The 'good milk'

On talking to the women about the methods, times, duration and frequency with which they breast-fed their children, I very soon came up against the question of the qualities of breast milk, considered indispensable for the infant's nutritional needs.

I already knew of the commonly held opinion that the mother's milk, given to her child from the very first, is the 'food' of the new born. 'Breast milk is "his" food, the food which makes the baby's body grow, *kuzenga mwili*', was the answer referring to the function of breast milk. The expression *kuzenga mwili* literally means 'to build the body', in the same way as to build a hut. In the women's minds, milk is thought to act in two closely connected directions: to build the body and to instil strength and force.

The demonstration of 'good breast milk', *mele maswanu*, is a child in good health, that is, a robust child, growing well. Breast feeding *per se* is therefore not sufficient; the mother's milk must have well-recognised characteristics which are able to guarantee the development and health of the child. Therefore, when the women talked to me about the requisites of 'good milk', they all defined the same in terms of *quality* and *quantity*, both of which are indispensable attributes to satisfy the appetite and the growing needs of the infant.

After whiteness, a characteristic of 'real milk', distinguishing it from the yellowish colour of colostrum, the women introduced the second quality, the 'consistency'. Milk must have the right degree of 'fluidness' and must not be like water. In this sense, the 'consistency' referred to in 'real milk' is not the consistency associated with colostrum.[25] Whiteness, *mele mazelu*, and consistency, *mele mazito*, do not exhaust the attributes of 'good milk'. It must also be sweet, *mele mzinzi*: 'whiteness', 'consistency' and 'sweetness' must be combined to give a 'good milk'. This synthesis is held together by a fourth characteristic, which seems to act as a bonding agent, as it is indispensable in obtaining the best result possible in the child's development: 'good milk' must have the 'right temperature', it must not be or become hot. I will come back to this later.

This combination of attributes is still not enough with regard to the quality of 'good milk'. In fact, the women also talk about the 'appropriate' quantity, confirmed by a satisfied and calm baby, who does not seek the breast excessively and who, consequently, does not cry insistently. According to the mothers, it is important that the baby is quiet and does not cry frequently, especially in the first months, so that the milk sucked goes 'to building his body, and is not lost in the crying' – as though to say: he must not waste energy uselessly.

The qualities described for 'good milk' are not everlasting constants. Not only do the child's needs, which change with his development, render the mother's milk insufficient for his food requirement, but also during the breast-feeding period, the women believe that situations and events may intervene and modify it, more or less significantly, with different degrees of effect on the baby's health.

We shall now consider these modifications, true alterations, and what they mean for Gogo mothers.

The 'good quantity' of milk

While Gogo women believe that milk may vary from one woman to another, all agree that it is impossible for there to be no milk after delivery, thereby agreeing with the opinion held by Western scholars (Jelliffe and Jelliffe, 1978; WHO, 1985; Mohrbacher and Stock, 1997). The confidence displayed by the women in believing milk to be an essential element in the mother-baby relationship must be the consequence of an empirical confirmation. On the other hand, the extraordinariness of such an event goes hand in hand with its rarity. In my two-and-a-half years' stay in the village, I never saw a case of a new mother without milk. Should such an event occur, the women would consider it, not as a part of nature, but of magic – and so strong a magic as to be without remedy.

A shortage of milk or the sudden disappearance of the same is another thing. The women, and especially the older ones, told me that there are bad people who are able to make milk disappear suddenly, or gradually diminish the quantity that the woman produces. 'Wicked people, *wasugusugu* or *walabalabala*, are able to make the breasts dry', they told me, adding that during her period, a sterile woman can 'dry up' a woman's breast, just by looking at her while she is breast feeding. The strength of this belief is such that if a woman thinks she has been 'glanced at', or been the object of a spell or curse, the mere fear of the truth of the event will make her milk go away.

I myself witnessed a case of presumed 'wicked eyes', which did in fact make the woman lose her milk, confirming the importance of emotional states on the let-down reflex mechanism. In this particular case, the baby was eight months old and the traditional cream of millet and water had already been introduced into his diet, and the mother got round the loss of her milk (caused, she believed, by a jealous woman) by increasing the number of daily meals. If the child had been very small, his future would probably have been more uncertain and difficult.

A woman may also experience a sensible reduction of her milk, until its complete disappearance, due to illness. In general, the women breast feed even when they are unwell, but should the indisposition turn into something more serious, a woman's milk may disappear or diminish. The relationship perceived by the women between a sickness causing loss of appetite and the reduction of milk is interesting, the only situation I saw in which food consumption was seen as related to the production of milk.

If the total disappearance of the milk flow is an infrequent event, the occurrence of a scarcity of milk due to the child's increasing crying was, the women told me, a different story. 'A baby who cries too much is a baby who is sucking milk which does not satisfy his hunger and make him grow', was the explanation offered by two traditional birth attendants when faced with continual crying, as they repeated with all the authority of their role the cry-hunger correlation that all the women trust when offering the child the breast.

If this presumed lack of milk occurs in the first months of the baby's life, first of all the woman can try to remedy it, as she is taught by the older women, by swallowing a porridge of millet flour diluted in water to which black pepper has been added, a substance which is believed to be useful in stimulating lactation. If the hoped-for results are not forthcoming, the mother may decide to go to the traditional healer, *mganga*, who will make some incisions on her breast, on which he will then spread some medicine able to increase the quantity of milk. Some healers also give roots, ground to a flour, to be mixed with food. If there are still no results, she will be more and more worried and will try other healers: if her role as a nurturer does not respond to her baby's needs, she feels she is putting his health at risk. Her tension, together with the additional food she is forced to give to her little one (which, in fact, reduces the number of breast-feeds during the day), produce a slowing down of the let-down reflex, convincing her more and more of her lack of milk. In this way, the mother unknowingly triggers a vicious circle, to the detriment of the psychophysiological process and dynamics typical of breast feeding.

If, however, the baby's dissatisfaction occurs around the age of three or four months, the mother, as we have seen, believes the scarcity of her milk to be due to the baby's increasing needs, and therefore considers it normal to add supplementary food to her own milk.

In this phase of starting up the breast-feeding process, the mother's milk may be considered 'scarce' also with regard to the quality of the same, in the sense of being 'watery', *mele mapelupelu*, light milk, milk without quality, so unnutritious as to motivate the precocious introduction of additional food to the baby's diet. This negative characteristic of a mother's milk is attributed to the woman's individual physical constitution, in the same way as that mother who showed off by saying that she fed her nine-month-old-child with only her 'good milk'. They are examples of good or bad 'fate'. I was told that 'Some women, because of their nature, are not able to satisfy the baby with just their own milk at a very early stage, while others are able to do so for longer'.

An inadequate milk for the nutritional needs of the infant is not assimilated as 'bad milk'. The latter, in the minds of the Gogo women, comes to be through special circumstances, all aimed at altering a milk which was previously 'good milk'.

The 'bad milk'

In order to fully understand the presence of 'bad milk' in the mother's breast, it is necessary to remember how much Gogo women believe that breast milk is a fluid which is extremely sensitive to those changes which take place in the woman's body, altering her *normal* equilibrium. The results of this alteration of her milk are always perceived as negative. 'It may happen that the milk is no longer the same; it is no longer the baby's "*own*" milk. It is altered and when it reaches his stomach it cause diarrhoea and sometimes, even vomiting', my informants underlined.

The women attribute change in the physiological equilibrium of the female body to events beyond their control, but it may also be a consequence of their behaviour, or the result of choosing, or not choosing, of negligence or of thoughtlessness. Among the events not attributable to the mother is a high fever. Even one, sudden 'fit' of high fever is sufficient to cause an anomalous heating of the woman's body, which in turn, heats her breast milk: '*Yatamwa, mele gaza galimoto*', [the woman] is sick, her breast milk is beginning to become hot, is the more recurrent expression with which women denoted a breast milk's alteration because of a fever.

This milk is no longer a good milk for the child: it becomes 'bad milk', *mele mabi*. In fact, should the child suck it, all the women unanimously concluded, he would very soon have diarrhoea and vomiting.[26] A mother knows that she must avoid giving that milk to her child, and so she squeezes it out and awaits the return of normal body temperature, before allowing the baby to suck again.

On the other hand, the alteration of the mother's milk may be attributed to her behaviour, ignoring the close connection between her milk and her own body. In this way, even a healthy mother may spoil her milk, for example if she exposes her breasts for too long to the sun's rays, due to the habit of going uncovered. One of my elderly informants told me: 'Like a fit of fever, the sun heats the milk, *mele gajemka*' (literally, 'it has heated', in the woman's breast, making it bad for the baby). In this case, the mother must apply cold compresses to her breast, and squeeze out the 'bad milk'.

Another case of 'bad milk' is milk which has become sour, *chachu* (Swahili word), because it has remained too long in the breast. This may happen when the woman spends the day working in the fields, or in the market, leaving her young child with an older brother or sister. The same thing may happen if the baby does not feed with the usual frequency because he is unwell, or refuses the breast for any reason. In these circumstances, it is also necessary to squeeze out the milk which has become *chachu*, 'sour milk', before feeding the child again.[27]

The circumstances described above do not consider the intervention of traditional healers, *mganga*, because Gogo women know how to read the signals of their own bodies and those of the child's. Should a woman not have squeezed out all the heated or sour milk, the child's diarrhoea will only be episodic, and will clear up once the 'bad milk' has been expelled. However, in her role as nurturer, it would be a sign of negligence on the mother's part, and she would try to get out of the situation, to avoid the other women's criticism.

The breast may also become swollen and painful for no apparent reason. The sudden appearance of these symptoms is, on its own, a sign of 'bad milk'. This is one of those circumstances in which a woman may decide to go to a healer for a remedy, above all if the baby is very small. The traditional healer, usually an expert on the subject, intervenes by making small vertical cuts across the chest, just above the woman's breast, and applying medicine, a powder obtained from herbs and leaves, or shoots. Each healer has his own pharmacopoeia, jealously guarded, even to the point of not revealing the places in which he collects his remedies.

'Bad milk' in only one breast

While moving freely around the village, spending my time with the women, going together with them to the vegetable gardens, the fields and water sources, on visits to friends and neighbours, I had already seen some of them breast feeding from only one breast, the right or the left. I had heard of this habit from my first meetings with the doctors in Dodoma, a habit described by them as one of the causes of malnutrition in infants fed exclusively with breast milk.[28]

In general, when I got close to the women, they hid their breasts, a gesture which I had at first attributed to embarrassment in front of a stranger. In these chance encounters, I had only noticed the lack of a repeated preference for one breast or the other, while I did encounter a certain reserve on the part of all the women about giving me information as to the reasons for such behaviour. I wondered whether the rebukes received or heard at the clinic for this practice, which the health personnel of the clinic and the white doctor there deemed damaging for the baby, contributed to their reserve.

Over and above my observations, I had heard, in various circumstances, of the eventuality of a mother having a 'bad breast', *itombo ikali* or *itombo ibi*,[29] and I had found myself faced with silence or vagueness which discouraged me from asking questions about the issue. It was clear that I would have to be patient when dealing with this aspect of breast feeding, and wait until I was considered decidedly outside the Gogo women's stereotypes of whites, *wazungu*.[30]

Here, as on other occasions, circumstances came to my aid.

One afternoon I was in a household near the high part of Muwuti, close to the hills, intent on talking with a small group of women about the different circumstances which they believed caused alterations in a mother's breast. The animated conversation continued when we saw a young woman with her small baby attached to her breast approaching. As her breast was uncovered, I noticed that the breast not being sucked by the baby in that moment was flaccid and dried up.

After the initial greetings, encouraged by the fact that the woman had not covered her breast in my presence, I tried to engage her in the conversation. In this way, she mentioned that her baby was growing well, but that he had 'been on his own for some time'. Surprised by her words, I asked her what she meant by 'his being on his own for some time' and she answered that the baby 'had been eating alone for some time'. Indirectly, she was trying to tell me that the baby had been left alone after the death of his twin. I did not insist anymore; African women are very reserved, or do not talk at all about dead babies. The woman's willingness to talk, however, moved me to ask her a direct question about her dried-up breast, and this gave rise to a conversation in which, surprisingly, all the other women present joined.

I thus learnt that, after birth, twins are put to the breast without any particular rule: chance establishes which breast is sucked by one twin, and which by the other. From that moment onwards, however, each newborn is given the breast which he first sucked, the one from which he received colostrum – that one, and *only* that one, is 'his' breast. Tradition teaches a mother not to exchange breasts between the two babies, not even if one of the two dies. The women's answer to the question as to their behaviour was invariably the same: 'We have been taught that we shouldn't mix, each baby has his own "good milk".'[31]

This conversation was opening the way to a deeper investigation on unilateral breast feeding. I sensed that the women who were present, directly or indirectly, knowingly or unknowingly, were allowing me to cross a boundary that until just a moment earlier, had been forbidden to me. I had to be careful not to offend them and not to give the impression, with my questions, of judging them, for a behaviour which they believed was a 'normal' response to determined events.

From that afternoon onwards, I was tacitly 'entitled' to know even this aspect of breast feeding, as though those women had wanted to test me on a subject which they knew was criticised by white people. Having passed the test, I was able to go deeper into the reasons behind such a practice. Besides the death of a twin, I was able to identify two orders of things which promoted unilateral breast feeding among Gogo women: (a) physical causes or causes seen as such; (b) causes attributable to events that could not be tested objectively.

The physical causes are ulcers, abscesses, rhagas (*ipu, donda, kidonda, mbadi*), which, as well as being very painful, also according to the women, make the milk of that breast go bad. On the other hand, that the milk is spoilt is proven when it comes out and mixes with pus, *ufila*, and blood, *sakami*, leaking from the lesions near or on the nipple.

Once healed, the woman does not always reactivate the breast, in spite of the recommendations of the health personnel, to whom she usually turns for treatment. These recommendations are at loggerheads with what the traditional healers affirm, and that is, that the milk from the sick breast is irremediably bad, even once healing has taken place, thereby proving that turning to Western medicine does not exclude the contemporaneous presence of the dictates of traditional medicine.

Alongside visible manifestations, there are others felt only by the mother, but which are not objectively verifiable. This happens when the woman feels apparently unmotivated persistent itching, or sudden sharp pain, in one breast only; these are, for the mother, symptoms indicating 'bad milk', which may not be sucked by the baby without causing him harm. In the former case, the itching is attributable to small animals, *vidudu*, while in the latter, the women speak about pain without specifying the reason.[32] When these symptoms arise, the woman immediately stops breast feeding from the afflicted breast, and if itching and pain continue for some time, that breast is rarely used again, even when the itching and the pain disappear. Sometimes, the woman continues to breast feed unilaterally not only when the symptoms have disappeared, but also after an eventual new birth.

If the decision not to allow the baby to suck from one particular breast is made by the mother following physical manifestations, or physical symptoms felt, there are two other cases, although lacking in symptoms and external signs, that are resolved by not using that breast any more. These represent the second order of causes, described above, attributable to events which cannot be objectively proved.

In the first case, although the woman does not feel any symptoms, the baby is afflicted with diarrhoea and vomiting. Faced with these manifestations, not infrequent during the child's development, the mother possess various remedies, as we shall see, according to the different causes considered from one episode to the other. If, however, what she does, as she has learnt from older women, does not have the desired effect, she begins to ask herself tacitly, if some kind of change in her milk has caused her little one's diarrhoea, resistant to her remedies.

She will start investigating by paying attention to her baby's reaction when he sucks from each breast. If this does not give her the answer, she will make a more direct test. This consists in her squeezing a little milk from each breast and keeping that from the left one separate from that from the right one, in two small shards from a cooking pot, used to cook *ugali*. She then places the earthenware shards on the fire and watches them, to see if the milk contained in both changes colour. If, in her opinion, there is a colour change in one of the two samples, then she will immediately stop giving milk from that particular breast.

Another possibility is to go to the healer, who will suck the breasts and decide the quality of the milk contained therein: 'sweet', *mzinzi*, 'bitter', *ikali*. Should the healer determine that one breast has 'bitter' milk, the woman must stop breast feeding with that breast, because it is that one that is causing diarrhoea and vomiting in her baby. There is no remedy to this situation.

The case of a baby who refuses a breast at a given moment is different. '*Mwenecho du yalema*' the women told me, to underline the baby's will in not wanting the milk from one of the breasts, the left or right one indifferently. This is a rather complex and mysterious case, the women state: not even the healer who is consulted can taste the difference in the milk from the left breast as opposed to the right. In his opinion, the taste is the same, without any taste at all, *mele ipolo*. Only the baby recognises a different taste and the mother must not force him in any way to use the breast he has refused. The women were always very vague when explaining the reasons for the infant's refusal, mostly repeating the healer's words: 'The choice is for the baby, only he knows, not even the *mganga* knows and cannot do anything about it'. Half-sentences, and winks, made me understand that this could be one of those cases in which milk is poisoned by an evil person, or a witch, who wants indirectly to damage the baby's mother.

Notes

1. In these statements, I found that sense of 'naturalness' which Geertz attributes to 'common sense', when, on enumerating its qualities, he puts it among the most fundamental one. The statement 'it's natural that it should be so', often used by the Gogo women with regard to breast feeding, refers to those 'intrinsic aspects of reality, the way in which things go' (translation from Italian) with which Geertz has interpreted the definition of the 'naturalness' of the 'common sense' (Geertz, 1988: 107).

2. In this sense, Pomata's article (1983), starting from the discouragement of the position of some female scholars' works on the incidence of the heredity of female biology on women's roles in certain societies, helps us to understand the complexity and, at the same time, the problematic of reflections, which dialectically call biology and culture into question. See also: Oakley (1985); Ardner (1993); Caplan (1993).

3. Anthropological and historical studies on breast feeding have shown that it is not just a physical (natural) process, that is, an expression of the physiology of the female body, but an integral part of the social and cultural organisation of a community. Anthropological literature is rich in this sense and will be frequently cited in the course of this work. At this point, I will refer to only a few among the many: Kitzinger (1980), Raphael and Davis (1985), Quandt (1985; 1986), Fildes (1986), Dettwyler (1986; 1988), Palmer (1990), Maher (1992), Riordan (1993), Stuart-Macadam and Dettwyler (1995), Thompson (1996), Dixon Whitaker (2000), Walker and Adam (2000).

4. This distinction previously appeared in Mabilia (1996a: 198).

5. Goksens's article (2002) on the action carried out by the interrelations between intentions, behaviour and social pressures on the promotion of breast feeding is interesting. The author, on studying the choices of Turkish mothers with regard to breast feeding, points out the incomplete or simplistic vision of Fishbein and Ajzen (1975) – theory of reasoned action (TRA): '*Our*

results point to the significance of instrumental support, social embeddedness and infor-
mationasl support as enabling factors that help translate intention into behaviour'
(2002: 1751).

6. At the time of my research, the village of Cigongwe was rather exceptional
   within the framework of the district, in as much as the percentage of women
   who went to the clinic, especially for the first delivery (the Ministry of Health
   had made the hospitalisation of primiparas compulsory) was about 25–27
   percent, compared to the district average of 3–4 percent. Such a high percentage was mainly due to the collaboration between the nursing staff and a
   traditional birth attendant, highly respected by the women in the village, and
   whose presence, in a culturally alien environment, gave continuity and a
   sense of security at such an important moment, still mainly stressed by traditional models.

7. Numerous medical publications have shown that an infant can regulate his
   need for milk thanks to the mechanisms of let-down, or milk-ejection, reflex
   (Jelliffe and Jelliffe, 1978, 1980; Helsing and Savage King, 1983; Glasier and
   McNeilly, 1990; Barry Lawrence, 1994; Quandt, 1995; Woolridge, 1995; Stuart-Macadam, 1995; Mohrbacher and Stock, 1997).

8. The custom of giving liquids or food, sometimes already chewed by the
   mother, before breast feeding is widespread in the traditions of other regions
   of Africa and the world, including Europe, as many studies testify (Jelliffe and
   Bennett, 1972; Knutsson and Mellbin, 1969; Fildes, 1986, 1995; Palmer,
   1990; Almedom, 1991a, 1991b). In my previous experience among the
   Imenti, one of the nine ethnic groups of the District of Meru, Kenya, I
   observed the mothers chew small portions of banana or yam, and give them
   to the newborn from the first days of life. Furthermore, tradition says that
   after the birth of the baby, the father gives the mother a gift of sugar cane,
   four if the child is a boy and three if a girl. These canes are chewed by the
   mother who gives the juice to the baby (Mabilia, 1991).

9. On the question of cattle, see note 11, Chapter 1.

10. There is a rich medical bibliography on the importance of colostrum in
    strengthening the baby's immune system. With reference to the benefits and
    properties of the same see Macy et al. (1953); Fomon (1974); Hooton (1991);
    Barness (1993a); Lawrence (1994); Cunningham (1995).

11. *Manhandu* does not necessarily mean the cream of boiled milk. *Ubaga* also
    forms *luuhandu* when it cools down.

12. Anthropological literature is full of testimony to different beliefs and behav-
    iour with regard to colostrum: from a practical attitude like that of the Gogo
    mothers, and of the Barbara (Mali) mothers, to a refusal to allow the baby to
    suck colostrum because not useful or even damaging, like that of the Balata
    women (Guinea Bissau), who wait 3–4 days before breast feeding, because
    they believe that the first milk is life-threatening for the newborn. The Fula
    and the Mandinga, on the other hand, believe it to be useful, because it gives
    strength and protection against illness to the newborn (Dettwyler, 1987;
    Gunnlaugsson and Einarsdottír, 1993).

13. The work of McKenna and Mosko (1993) underlines the activity of the child
    during sleep: by positioning and repositioning himself relative to the mother's
    position, he seeks the best position for feeding.

14. The supposed strength exercised by an infant's cry in bringing out maternal
    milk has been recorded in other parts of the world than sub-Saharan Africa.
    For example, Raphael and Davis (1985) find it among the Igorot, a population

in the mountains of the Philippines. It is also true that medical science confirms it, when it recognises activation of the let-down reflex as a mere response to the desire or thought of the mother to breast-feed her child, 'conditioned stimulus' (Jelliffe and Jelliffe, 1978: 22). See also the study on the effect of cry stimulus carried out among the primipara in Stockholm (Lind et al., 1971: 293).

15. This way of thinking does not diverge from the dictates of biomedicine when it affirms that the introduction of food, together with mother's milk, is necessary after the first six months of exclusive breast feeding, for the growing needs of the infant (Jelliffe and Jelliffe, 1978; WHO, 1985; WHO-UNICEF-USAID, 1992; Mohrbacher and Stock, 1997; UNICEF, 1991, 1999, 2002).

16. These Cigogo expressions indicate the evaluations made by Gogo women with regard to the various degrees of child development: by *yakali mdodo* they mean a very small child, about one or two months; *yalasuga* is a child of three or four months and *yalawahapa* is a grown child of about eight or nine months. These various phases of growth must be understood as referring to the growing need for food.

    Nhili, together with *ubaga* and *matili*, or the Swahili term *uji*, frequently used today, above all by the younger women, is a cream of millet and water, the first food used together with mother's milk. U*gali* is porridge made from millet and sorghum, accompanied by *ilende*, a green-leaf vegetable, the favourite of the Wagogo of all the vegetables available with the first rains. This combination of foods synthesises the concept of what a meal is for the Wagogo.

17. The previously cited Raphael and Davis (1985) supports the belief in the correlation between the child's frequent crying and the inability of the mother's milk to satisfy its nutritional needs. See also La Fontaine (1981).

18. For adults, millet porridge is cooked in a bigger saucepan, *nyungu*.

19. Cow's milk is considered special food. The Wagogo do not use fresh milk, *mele masusu* – soured milk is preferred. The idea of boiling milk is foreign to them, in that they believe that it thereby 'loses strength', is impoverished.

    They distinguish *mele masuce*, whipped milk deprived of fat, from *mele mphopota*, whole milk. The latter is simply a milk which 'has slept', as the Wagogo say, that is, it has remained in its container for a night, sufficient to sour. The two kinds of milk are normally smoked using different types of bushes, which give them a particular taste. Milk is a food to be eaten and not drunk. It is consumed with porridge.

20. The advantages of germinated millet for the nutrition of children during weaning is known (Brandtzaeg et al., 1981; Mosha and Svanberg, 1990; Savage King and Burgess, 1993).

21. Here we are faced with the clear distinction between weaning intended as a process, and a weaning as a complete termination of breast feeding (Raphael, 1984; Almedom, 1991a, b).

22. All the mothers have the children's Growth Monitoring Card with which they should go to the clinic each month to check the child's weight. As far as possible, relative to their agricultural work and the weather, the women are diligent in keeping this monthly appointment. However, they do not always pay much attention to the information registered in the card, preferring to evaluate the child's age according to those criteria used by their mothers before them, in assessing their offspring.

23. The impracticability of breast feeding in a pregnant woman will be examined in the next chapter.

24. These tiny insects of which they spoke were called *manoonoo* or *mangh'ongh'o*, but I was unable to find out exactly what they were referring to.

25. The use of adjectives in Cigogo is rather complex in as much as they assume meaning, sometimes an antithetic meaning or shades of meaning, only in relation to the context in which they are used. The case of the counter-position between 'dense' and 'liquid' milk is a good example. The 'density' of colostrum assumes a discriminate value when applied to *real* milk which is more liquid, but when the latter is compared to low-quality milk, then the same attribute assumes a positive value.

26. Diarrhoea and vomiting are the clear-cut signs that something is wrong with the mother's milk, although, as we shall see, it is above all their persistence which makes the woman worry about the quality of her milk.

27. There is a vast anthropological literature on beliefs relative to possible alterations of breast milk. I remember, for example, the Khmir women of northwest Tunisia, who believe, as do Gogo women, that milk heats up as a result of the body's overheating. In general, this occurs when the woman has been engaged in heavy work, such as collecting wood in the mountains. Unlike the Gogo beliefs, however, which dictate that the spoiled milk must be squeezed out, here it is sufficient to wait for half an hour to have good milk again (Creyton, 1992: 45).

    Bambara women believe that the milk becomes bad when it stays too long in the breast. If the baby sucks this milk, which is considered 'old' , diarrhoea, and sometimes vomiting, will ensue. The women connect this change to the origin of breast milk. In fact, they believe that the milk is produced by their blood, and that once it reaches the breast, it must be sucked immediately, otherwise it becomes bad (Dettwyler, 1987: 638).

    Outside sub-Saharan Africa, we find, for example, the Igorot women of the Philippines, in the previously cited study be Raphael and Davis, who, on returning from the fields, wet their breasts with cold water, because should the baby suck from a hot breast he will have diarrhoea (Raphael and Davis, 1985: 35).

28. An investigation which I carried out into the health of children below five years of age, taken to the clinic for monthly weight controls, proved this belief to be inexact. The data relative to health status in children below two years, breast-fed from a single breast, revealed 9.1 percent of cases of serious malnutrition, compared to 9.3 percent in children fed from both breasts. If unilateral breast feeding does contribute to low health status in children, it is certainly not the only reason. The data collected at the clinic indicated that 5.4 percent of mothers breast fed from a single breast, much lower than the 14.9 percent resulting from the sample which I used in the village. The reason is to be attributed to the fact that the women who attend the clinic always try to hide their breasts, fearing rebukes from the health personnel. The justifications given for unilateral breast feeding by the women in my sample were: five due to abcesses or wounds of the nipple, three due to itching, five due to pain, three due to the infant's choice and one due to the death of a twin.

29. The adjective -*kali* means bitter, poisonous, sour, as opposed to -*polo*, not bitter, not poisonous, sweet. Within these two categories, the Wagogo classify plants and bushes, *mukali* and *mupolo* (Rigby, 1966a: 9). Most medicines, *miti*, and the ritual sticks for lighting the first fire in a new dwelling, belong to the second category. Outside the classifying context of plants and bushes, the adjective -*polo* can also take on the meaning of tasteless, as opposed to *mzinzi*, sweet.

30. Remember that the village women used two stereotypes indicate a white woman: 'sister' (a nun) and 'daktari' (a doctor). See Chapter 1, note 12.
31. The concept of 'not mixing' will come up again in other contexts.
32. *Bobo* women in Burkina Faso recognise an illness which afflicts a woman's breasts and is caused by the closure of the mammary ducts due to small whitish worms, the origin of which they do not know. The presence of these animals impedes breast feeding and, in general, the women turn to a female healer specialised in the treatment of this disease (Alfieri and Taverne, 2000).

CHAPTER 4

THE 'GOOD MOTHER', THE 'BAD MOTHER': DIARRHOEA AS A SIGN OF SOCIAL DISORDER

Inside the problem

While paying more and more attention to the mothers' behaviour in various moments of the day, and listening with ever-increasing interest to any subject which seemed to be connected to their children, a case of diarrhoea in a three-to four-month-old baby gave a decided shift to my research. This fortuitous event would, in fact, lead me beyond the answer given to me on breast feeding as the 'natural' (obvious?, usual?) response to the newborn baby's nutritional needs.

After an intense day spent with a number of families, I was making my way to the clinic, alone and on foot. My assistant had stopped off at friends' in the hope of buying some cheap millet and charcoal. We intended to return to town that evening. I was walking at a steady pace through the brushwood along the path, when I saw a woman coming towards me. I remember that it made a strange impression on me because the sun, by now low on the horizon, allowed me to glimpse her minute shape only through a sparkle of trembling lights which distorted the contours of my surroundings. I was even more surprised when, on focusing on the woman, I recognised her immediately. I remembered her well because of her reserved and even reticent behaviour. I had seen her, together with other women, at the river or in one of the hamlets, during those relaxing moments that coincide with the hottest hours of the day. I had noticed that she always kept apart from the others and very rarely spoke, refusing to participate in even the most animated conversations. I was, therefore, very curious about her sudden appearance. On meeting, she hurried the usual greetings and asked me to help her, but secretly. She must have noticed my surprise, as she continued to beg me to accompany her by car, the following day, to a local healer, at some distance from the village. Her child had been very ill with diarrhoea for some days and all the remedies she had tried had been in vain. He was so weak that he was not sucking anymore. I understood that only her anxiety for her child's health had given her the courage to ask for my help.

I gave up the idea of returning to town and agreed to accompany her. She gave me instructions about meeting her the following morning, 'before the sky lightens', in a special place, 'where we will not be seen', she told me conspiratorially. I had serious doubts about the latter condition. I had always had the impression that the whole village knew of my whereabouts, in every moment! Keeping these thoughts to myself, I said goodbye, restlessly thinking about what awaited me the next day.

The woman was already waiting for me when I reached the agreed meeting place at the agreed time. She hurriedly greeted me as she got into the car with her child in her arms, bundled up in a *kanga*, and sitting herself beside me she told me where to go. From that moment onwards, she did not say a word, except to indicate the track or path to take. An hour later, we had not met a soul and the road was becoming more and more uneven and rough. We were driving among brambles, bushes and brushwood, every now and again avoiding acacias and baobabs, when she told me to descend towards the left onto the dry bed of what must have been a wide river, but which was, at that time, just a sandy track, dotted here and there with holes for collecting water, a sign that there was a village somewhere around. The descent on the riverbed worried me, but the ascent a little way further proved to be even more difficult and I was forced to make more than one attempt to find a relatively easy way of making it to the top.

Once on the level, she indicated the direction across a sun-burnt savannah, enclosed by hills in the distance. I took advantage of this break in her silence to ask her how the child was. He was still sleeping on her lap and after a few minutes of silence that seemed eternal, she began to speak. Her anxiety for the child's health and the unease that had been tormenting her for days was evident from her speech, sometimes clear and sometimes confused. I was surprised by her loquacity and became more attentive, understanding, from the flow of words, the woman's desire to reexamine, for what must have been the umpteenth time, what had happened to her. It was as though she was reviewing the preceding days, minutely examining and reexamining her behaviour, in order to understand what had happened. She repeatedly used the words 'behaviour', 'breast feeding', 'elder women', 'taboos', 'sex', 'men', all mixed up with 'tradition'.

In a moment of silence, I asked her what could have happened to her child and she answered in a rush:

> I have been a good mother; I breast fed, I was careful to do the 'right' things, I haven't slept with my husband in all these months. Then why does the baby continue to have diarrhoea, to become so weak that he can't even suck my milk?

I understood that the question was not addressed to me, certainly she had other interlocutors.

In thus giving vent to her feelings, dictated by the anguish and tension accumulated in the preceding days over her child's health, this mother was giving body to the allusions, to the half-sentences, to the saying and not saying with which some women had allowed me to glimpse, or only to sense, a scenario in which the sexuality of a woman, mother and nurturer was subject to rules which, if not observed, could have serious consequences for the baby's health.

Her agitated way of speaking had also revealed, as well as her unrest about her child's health, the worry about her image as a mother and about the consequences on the network of relationships, on the quality of the ties uniting her, as a wife and a mother, to her family and community environment. After all, she had, in that brief dialogue, defined an articulate image of a woman – wife, mother and nurturer – projected into the typical dynamics of conjugal relations, family ties and gender relationships.

From that journey onwards, I could no longer escape the problem: in order to understand the vicissitudes of breast feeding, the way it reflects on the baby's health and on the mother's role, I would have to face themes connected to sexuality, a complex field where not only rules, moral dictates and prohibitions, but also intimate ways of feeling and experimenting the body and its drives, permeate, harmonise and interfere with each other.

As on other occasions, I tried to enter this 'minefield' gradually, with great caution, feeling my way intuitively for the best road to take. I glimpsed two: the first, concentrating on the care and attention that a mother shows to her child during breast feeding, would lead me to investigate the illnesses which could slow down the child's development and bring him to those serious conditions of health that my interlocutor had so objectively and eloquently described. The second would lead me to investigate the sense of cohabitation, overlapping, and even conflict, for a woman in her roles as mother and wife, both of which are invested with rights and duties, but also with expectations on what it means to be a 'good wife' and a 'good mother'.

Motherhood and mothering

At this point in my research, it was clear to me just how much a woman was directed, right from infancy, towards an image of herself as an adult, where feeding and caring for children, her *own* children, were tasks entrusted to her. It is thus that the personality of the female child is constructed, through a process of identification with her mother, thanks to the continual sharing, firstly as a mere observer and imitator and, secondly, as a pupil and helper, of those practical tasks which then gradually become *her* tasks.

Not only does she see her mother surrounded by children, her own brothers and sisters, looking after them, breast feeding them and preparing their meals, but she also senses, by listening in on the women's conversations, that the mother's duty towards her offspring is, above all, to maintain her children in good health. It is in this way, that she senses her mother's anguish when one of her children is affected by an illness and learns, growing up, just how many dangers lie on a child's path. The image of a woman to whom many tasks are entrusted gradually takes shape in her mind – 'motherhood' and 'mothering' assume characteristics and values which she will interiorise so deeply that they will be intrinsically tied to her personality and to her individuality as a woman.[1]

As she well knows, marriage signifies deep changes for a young woman, starting from her status as a married person, which carries rights, but also very well-defined duties and responsibilities. The fact of going to live with her husband's family is, on its own, a change which reflects on, and becomes part of, her way of being, of her most intimate experiences and which inevitably interacts with her personality.

As the months go by, a young woman knows that the women of the *kaya*, and in particular, her husband's mother, are discreetly observing her, *silently*, but more and more attentively, in order to catch the first external signs of pregnancy in her body.[2] On the other hand, she herself anxiously awaits those changes *inside herself*, which she has learnt are premonitions of a soon-to-be motherhood and which will confirm her ability to reproduce. Pregnancy, and the consequent maternity, assumes a double value. In fact, in her eyes, if giving children to her husband means that she is fulfilling the first duty of a married woman, it also confers, at the same time, the full status of an adult woman, which is characterised, as we have seen, by the formation, with her offspring and the cattle assigned to her, of an economically independent group within the domestic group, the *nyumba* within the *kaya*.[3]

The women count the months of pregnancy, following the lunar phases, starting from the first month without menstruation: the first month has passed when the moon presents in the same phase as the one in which the woman recorded her state of amenorrhoea. The older women remarked, with a certain sense of pride, that it was not necessary to go to school, or to possess a calendar,[4] in order to know the process of pregnancy, because 'it's counted in the head'.

The period of gestation lasts for ten moons and is experienced by the women as a normal period within their reproductive cycle, so much so that it does not require particular attention, neither with regard to the rhythm of work, nor to the assumption of food.[5] The people around the pregnant woman do not have to pay particular attention to her state. To show interest or curiosity, for example by asking how many months until delivery, or how many months have passed since the 'end of menstruation', makes the woman suspicious and agitated, thinking that she is the object of some kind of malice or spell, to the degree that she may ask herself: 'Why so much interest? What do they want to do with my pregnancy? Why do they want to know?'

When, just as inopportunely as unknowingly, I asked a young woman how far on was her pregnancy, on seeing her weeding millet beneath a scorching sun, she looked at me with both surprise and disconcertion. After a few moments of reciprocal embarrassment, she must have thought that my rudeness was due to ignorance as she answered: 'A future mother must never draw attention to her state, she must protect her pregnancy from bad or envious people,[6] therefore not to talk about it is to safeguard it.' On the same occasion, I learnt that a woman can speak about it with her own mother, with her elder sister or with her mother-in-law, although with regard to the latter I received, at a later date, contradictory replies. In any case, not even the women closest to her may mention her pregnancy in the early stages, *muda munyanhwi* (literally, it is still young).

If the woman's daily routine continues without particular changes, she must, however, be careful from the very beginning of pregnancy. The husband is also involved in this task, as it is a common belief that sexual activity is useful for the growth of the foetus, if traditional rules have been well observed. 'The baby is also the result of the mother's and father's sexual work. Together, the husband and wife must look after the pregnancy, *yakudima muda*, as the rain which is falling makes the seed grow', the women told me. This 'looking after the pregnancy' with sexual activity, where the husband's sperm[7] mixed with the woman's blood, nourishes and helps the future baby to grow, must not, however, exceed the first three or four months, maybe even five. Once this period is over – the baby is complete and has grown enough – the woman must avoid sex, as her blood and colostrum are all that the foetus needs to grow well. I often received contradictory answers about the issue of the presence of colostrum at the beginning of pregnancy as a nutritional substance for the baby who is 'in the mother's stomach'. It is, anyway, a fairly widespread belief that was repeated to me: '*Goyakongha munda mpaka alalawila kuzelu*', literally, 'he sucks inside (the mother's) stomach, until he comes out into the light'.

If sexual relations continue for longer than is foreseen and allowed, the baby will be born dirty, *mchafu*,[8] covered with a nondefinable substance, a source of great embarrassment for the mother. The women told me: 'The baby is born grey, *mvutuku*, and the women who have helped the mother to deliver refuse to wash it. It must be done by the mother-in-law, *mkwe*.' When such a birth occurs, the mother is ashamed, *cevu*, because she knows the women think that she has privileged her husband's pleasure, rather than taking care of the pregnancy, as her mother and the older women taught her during puberty rites.

The tacit reproof that the women present at the birth show towards the new mother and the shame that she feels, of which I had confirmation many times, indicate the difference in the consideration shown to men and women, as only the latter are subject to disapproval for a behaviour which involves both sexes to the same degree.

This is not the only example indicating the relevance of the gender issue. The different roles that a woman takes on with her social maturity – becoming a wife, a daughter-in-law and sometimes almost immediately a co-wife and shortly after a mother – may be in conflict with each other in certain moments, due to imperatives that do not easily fit together (on the purely personal level as a woman), and from which a woman must be able to extricate herself.[9] In this way, the fullness of adult life is confirmed in the mind of a young woman, a condition that requires her to be not only constantly adaptable and flexible, but also firm and determined, as we will see, so that she may strategically manage her different and often conflicting obligations.

For Gogo women, maternity is a special event which significantly marks their existence, as women, as wives and as mothers, starting from the commitment to look after and guarantee the strong and healthy growth of their children. It is an important task: the woman is responsible for herself and for the life of another human being. This passage, defined as *matrescence* by Dana Raphael, is of great relevance, so much so that in many societies it is evidenced by rites, aimed at helping the woman to begin this new period of her life with a new and diverse awareness.[10] Washing the mother after delivery, the offer of a special soup, help from the same women who, after having assisted her in bringing her child into the world, stay with her for another week, are all events which are part of a rituality to help her through the passage to a new period in her life.

In fact, with the birth of her child, a woman, wife and now also mother, begins a long journey in which *mothering* means, above all, privileging the relationship with the newborn, a relationship which is conditioned by the child's needs and necessities, starting from those nutritional ones which are aimed at guaranteeing his development, health, protection and safety, in one word, survival. With this aim in mind, the long period of being a nurturer is distinguished by rules which condition specific behaviour and specific avoidance, to avoid compromising the most precious thing she has for the health of her child: her milk. It should not be forgotten that the tight bond which is thought to exist between a woman's body and her milk, a bond which encloses nurturer-milk-newborn, is a triangle which is supported in an equilibrium of fundamental importance for the good outcome of breast feeding.[11] 'Milk lives with the woman's body', a young mother told me, indicating simply, but effectively, just how much this fluid, produced by the female body, is considered vital and reactive to any change in her body. A mother, above all in the first months after the birth of her baby, has a very intense relationship with him,[12] not only in feeding him, but also in covering him with attention and affection. I often witnessed with what pleasure and care a newborn baby is treated by its mother. This can be seen in the careful and joyful way in which she touches and caresses the child's body, for example, when giving the child his daily bath, in playing with him, singing rhymes or vocalising; and, in the pleasure, as a few mothers told me, of being so close to him when breast feeding.

Furthermore, with physical care of the child, bringing up one's offspring also means stimulating and guiding his development as a person. A mother does not limit herself to being a nurturer, or to just helping her child to grow healthy and strong, but carries out the essential role, well past the breast-feeding stage, of directing the child towards relationships, teaching him about behavioural models, so that, as he grows, he may find himself able to participate in the life of the community.

It therefore seems clear that rearing a child is a very important role of responsibility for a woman, a role that, directly and indirectly, is watched over by the whole community. It is not by chance that the verb *kudima* is used not only in the expression 'to look after pregnancy', *kudima muda*, but also in the expression *kudima mwana*, to look after children, and in the expression *kudima ng'ombe*, to take the herds to pasture. Cows and children are the Wagogo's two most precious possessions, to which they dedicate great care. Therefore, during breast feeding, when the child is closest to, and most dependent on the mother, the woman is 'under special observation'. Her husband's mother watches over her, as do the women of the *kaya*, the neighbourhood and the entire community. This supervision is heightened whenever the child's growth is compromised by any symptom that may slow down, or put at risk, what is considered to be his normal process of growth.

The health of the newborn: a challenge to survival

The days immediately following birth and up to the fall of the umbilical cord, *mpaka nyumba yele*, represent a very delicate period of time, *mcisaga*, for the physical well-being of the newborn baby.[13] '*Yakali yalimoto, cigoli*', the women say when they want to highlight, with this expression, that a new born child, *cigoli*, is hot, *yalomoto*,[14] which means, particularly vulnerable.

Birth and the beginning of life outside the womb are conceived of being subject to a series of events, which can direct the newborn baby and those surrounding him, and even the area in which he lives, along different routes.[15] It is therefore necessary to pay attention to behaving oneself according to the rules. I was told: 'There are important prohibitions, *yena mwiko mu'waha*, to observe so as not to damage the health of the newborn baby.' These force the mother to remain inside the *ikumbo*, not to receive visitors, especially males, and not to let her child be touched by women still of reproductive age, whereas the older women, no longer fertile, may do so.

On the other hand, immediately after the birth, the father, as has already been mentioned, must seek out a diviner from whom he must obtain medicines and amulets which will offer a protective barrier around his small child against everything that could 'change his good wind', *mbeho swanu*, and therefore his 'good condition', both physical and ritual.

After this act, however, the task of maintaining the child in good health is the mother's exclusively, although, as we shall see further on, the father must respect a specific obligation, the omission of which could have tragic consequences for his child's life.

When the stub of the umbilical cord has fallen, around seven days after birth, the mother decides, *kulavya*, to take the baby outside the hut. She must do it carefully, preferably starting at sunset and for just a few minutes, covering the baby with care so that he may get used to the environment a little at a time. The mothers say: 'You must be careful, it is important for the baby to "take the air well", *hewa*[16] *swanu*, it is important that he "has a good wind", *mbeho swanu*.'[17] These latter expressions indicate worry, not so much about the atmospheric conditions (although attention is paid to these also), but about a 'good ritual state', intended as the physical, ritual and relational conditions which guarantee the newborn baby's correct entry into the community.[18]

If behaviour immediately following birth is part of a rituality intended to introduce the newborn baby into a new phase of life, taking care that 'everything is as it should be' the women know very well, however, that sometimes, even though everything has been done according to the rules, a baby may become ill from the first days of his life, from the very first moments, if that is his destiny: '*Yunji du zono yakulelwa, nyumba ikwela hodu, hone mwili wakwe hone mchafu*[19] *ya kutaamwa*', 'some (babies) immediately after birth, with the stub of the umbilical cord still attached, become ill, if their bodies are dirty'; and again, '*Watya yecewela,*[20] *mwana kecewela yakali du mgalika, alu mwili wakwe wapwituka*', 'the baby is ill, the baby has been ill from the start, his body has heated', my informants told me, with ill-concealed resignation with the bad fortune, *ngwamusanga,*[21] which seems to accompany some newborn babies from their first hours of life.

The same sense of destiny made the women repeat that there is no specific time to become ill, it can happen in any moment, outlining the life of human beings as a route, where periods of health alternate with periods of illness in the life of each one, from birth to death. Every human being, therefore, is born 'with his own illnesses, with his own death'. In this sense, to assert, as some women do, that certain illnesses are God's will, *Muwaha Mulungu*, means considering them in the same way as those natural events which occur in the life of every human being are considered.[22]

Nevertheless, they generally hold that there are two conditions necessary for the onset of illness. The first is given by an unbalance created between the body and the external environment, meaning external natural and/or ritual surroundings; the second is the result of the *agreement* between blood and the illness itself. One woman expressed herself as follows: '*Alu du nayo hone sakami yakwe vyeluvya hodu lyamkolela du*', underlining the propensity of a person's blood (*sakami yakwe*) to be receptive, predisposed (*kiluvya*, literally to be in agreement with) towards the illness, which then arrives, *hodu lyamkolela du.*[23]

A mother, on taking care of her newborn baby and in keeping to the letter of what she has been taught by the older women, is just as aware, because of these premises, how much the life of her child, in the first years of life, can be subject to many types of illnesses, in spite of her care and attention. The women therefore know very well that their expectations as mothers interact with the physiological processes of their children's growth, where health and illness are interrelated with the physical conditions and the imponderable presence of destiny, in an environment in which infant mortality is part of daily life.[24]

These considerations, however, do not lead to resignation. If keeping one's child in good health is a daily challenge, the women do not flinch from it, as they are well aware that the growth and health of their little ones also depend on their care and attention, on what they do or do not do.

A mother is the person who, more than any other, knows her child's needs and is able to evaluate his development, to perceive the signals of any change in his physical conditions, because she lives beside him, day and night, so much so as to know 'the rhythm of his breath', as one older woman told me.

> It is the mother who takes care of her baby: she is the one who carries him tied to her back during the day and who sleeps with him at night. Who can know him better? Who can better understand his needs and moods?
>
> It is the mother who sees everything, who caresses him, who feels the heat of his body, who notices when his mouth is hot, *mlomo ulimoto*.[25]
>
> How can you perceive the change in a baby's body temperature? The baby breathes near the mother, *ukumwehlela*, so she gets know the change in his temperature. She is always in contact with him, she takes him in her arms. The father never keeps him near. How can he notice the changes in his body?
>
> The father is never with his child and even when the child goes towards him, he only keeps him there for a few seconds, how can he know if he is well or not?

These were the answers the women gave, when I sought information about who looks after the children.

The image of a child that grows well, without problems, was outlined for me by the mothers through the child's specific behaviour: he desires the breast, he does not cry frequently, he sleeps well during the night, his body temperature is *right* and, as the months go by, he shows a growing liveliness towards his surroundings. The same progress made on a motor level – sitting on the ground, crawling, maintaining an erect position, taking the first steps – are all indicators, one after another, of correct development.

When, on the contrary, the baby refuses the breast but cries frequently, is not active during the day but is restless at night, these are signs indicating that something is changing in the baby's physical state and perhaps, announcing the onset of sickness. If his body becomes hot, *mwili wapya*, then he is really ill.

The mothers use the expressions *mwana mpolo* (literally, a sweet baby) and *mwana mkali* (literally, a bitter baby),[26] to indicate, in the first case, a baby growing well and without problems and, in the second, a baby that does not grow as he should, that gives problems. With the same expressions, they also indicate, in order, a baby that does not cry very much and one that cries frequently – both are considerations that are not very different from the previous meaning.[27] An older woman, referring to the onset of illness, told me:

There are many different illnesses, *aina ya nhamwa nyinji.* An illness is inside the baby's body: I can see his body give a start, *nghuwona mwili wakwe cikutinuka,* and I think, he is getting ill, *cikumwalabatya.*[28] You have to guess, the baby doesn't know how to speak.

As in most parts of sub-Saharan Africa,[29] for the Wogogo too, the attention, care and nutritional practices that a baby receives in the family environment, first of all from his mother, play a very important role in increasing his chances of survival.

To observe a mother in the daily care that she reserves for her baby means to consider an entire series of reactions, or absence of reactions, of activities, or absence of activities, of timeliness or delays, which she gives rise to, in maintaining or in restoring the baby's health. In her actions, or lack of action, cultural, social and psychological instances are woven together, determining the quality of her response or nonresponse. When a mother underlines the necessity of 'hazarding a guess', when she observes changes in the physical state of her baby, the importance of what *she thinks about* 'being well' or 'being sick' in someone who 'doesn't know how to speak' is evident. It is what she has learnt from her mother and from the older women, it is her fund of knowledge and beliefs about health and illness which determine her behaviour. It is her way of perceiving the gravity of this or that affliction which will decide, or not, her seeking out a specialist, *mganga.*

Considering that her resorting to a specialist's treatment is usually preceded by domestic intervention, on asking the women whether or not they go to a healer without seeking their husbands' permission, they invariably answered with an eloquent '*cinyele?*', that is to say 'secretly?'![30] It is the baby's father, *mnyamwana* (literally, the possessor of the baby), the one who will have to pay for the medicine, who will decide whether or not to resort to an expert, and in this sense, the women described, more or less explicitly, the strategies they used in attracting their husbands' attention, involving, in the most difficult cases, their mother-in-law. Furthermore, a mother may ask the latter to go and 'hoe the medicine', if the husband's return is delayed enough to compromise the baby's health. Some of the older women pointed out, with a certain amount of animosity, that when they were young mothers, the father of their child did not stay away from the village for more than two days. He had to guarantee his presence, his availability, in case a decision was called for with regard to his child's health.[31]

By trying to clarify some situations in which very serious forms of diarrhoea could put a baby's life at risk, I managed to delineate a general picture in which the women distinguish 'true illnesses', *mphungo,* from 'non illnesses', and both of these from those alterations in the baby's physical condition which may become serious,[32] but which are never defined with the term *mphungo.*[33]

A rise in body temperature or episodes of diarrhoea are among the most important symptoms of any change in the physical conditions of a baby. However, while the mothers always indicate a fever, *homa*,[34] as a state of illness, diarrhoea, *lwito* or *ida*, can assume, when a baby is afflicted, a very wide range of evaluations and considerations.

True illnesses, mphungo

During the first months of a baby's life, the mother believes that the fontanelle, *indosi* or *idosi*,[35] is a source of trouble for the baby and, at the same time, an important point to be observed when evaluating the danger of an alteration in body temperature. If it pulsates, all is well, but if it assumes a concave shape, 'the fontanelle has fallen', the women say, or, if it is convex or becomes tense and no longer pulsates, it means that the baby has a very high fever, his body is burning, *ciwili*[36] *chapya*. They very much fear this kind of condition. If a stiff neck and a tremor of the head also accompany it, these symptoms predict that the baby has *indosi*, a very serious true illness: 'Some fevers', a healer told me, 'are not easily treatable. This is generally true for very high fevers. When the fever lasts for more than two days, then it is a very serious thing.' The women, therefore, think that it is a good thing if the fontanelle closes early: a closed fontanelle does not cause any problems. They believe that water is useful to close it more rapidly, *lidosi lidinda mbela*. The mother gives the baby a few mouthfuls of water from the palm of her hand during his daily bath, also convinced that he wants it, as he opens his mouth along the mother's hand as she washes him.

This is not the only reason for which a mother offers mouthfuls of water to her baby from his very first months. As well as believing it to be good for health, the baby must learn to know it, an idea that we have already seen with regard to food, and by drinking a few mouthfuls, he will hear earlier as it opens his ears more rapidly. Not all mothers believe this to be true, especially the young ones, but even those who believe it to be a well-founded traditional teaching, agree that a baby does not remain deaf if he does not drink water. The older women state: 'Water opens children's ears so that they hear earlier, *makutu yatuje mbela, yehulika*. If you don't give water, he will hear later.'

The fontanelle, therefore, according to the mothers, is a very delicate and sensitive part of the baby's body, a part of the body which *can speak* about his physical conditions. It is not rare to see small pieces of wood or small beads tied to the fontanelle of the newborn baby – tied together with a narrow ribbon of skin laced under the baby's chin – or a spread of greenish paste,[37] as amulets or remedies as treatment or to safeguard health.

From the way the women described another true illness, with a very high fever, to me – cold arms and legs, while the trunk is very hot at waist level – it would seem to be malaria, particularly as such a state tends to alternate moments in which the body temperature returns to normal, *mwili wazapola*.

A similar case of high fever during the night, which disappeared at dawn, persuaded the parents to take the baby to the *mganga*. The latter, after having consulted the board, *akutowa cisanga*,[38] told the man and the woman that the fever was due to a discussion using strong words and threats, *mphimbi*, which had probably passed between the two, *mwagombana na yiunji*.[39] In telling me about this episode, which had occurred just a few days before in a nearby *kaya*, the women used the expression *'akwembelwa na laho'*,[40] which means that the baby became ill *due to something*, where this *something* indicated, in this particular context, a dispute between the baby's parents.

If the diagnosis is exact, the remedies offered by the specialist, medicines to be given to the baby to drink and other medicines to be put in the water used for washing him, will be effective.

When the fever is accompanied by a change in the colour of the baby's body that becomes yellow, then one is in the presence of a true illness, called *nyawana*. It only affects very small babies, *nyawana ya cidodododo*, during the flowering season of certain grasses, *kubongh'oka mahanze*, which have yellow flowers and are found in the undergrowth, *maluwa njano*, at the beginning of *ifuku*, the rainy season, between the end of November and the first days of December. As well as changing colour, the fever and the presence of yellowish faeces, the baby's lips are always dry, so much so that his continuous attempt to wet them is one of the characteristics of the disease.

On being faced with this pathology, the women all agree in saying that hospital treatment is not efficacious: traditional medicines are necessary, *wacisacila miti ye cigogo*.

> In the hospital they treat him with hanging water, *wawamtumbucila malenga*, they also give him blood[41] and he gets better. But, once he is home again, the baby becomes sick again, everything was done for nothing. So, we look for medicines, *hodu wawasola, wawhimba miti ye nyawana*.

They then have to search for *manyawana*, a herb from which the illness takes its name, for the preparation of fumigations,[42] and give the baby a broth of chicken mixed with special medicines supplied by the specialist. Only in this way will the child get better.

Ideje or *idejedeje*[43] is a very serious illness, *mphungo mbaha*, which instils fear. Change in body temperature is accompanied by sudden, and sometimes violent convulsions, which disappear just as suddenly as they come, leaving the child's eyes 'wide open'. The women say that these convulsions may return in time, without the fever.

Ideje may affect a child at any time: a few days after birth, when he is very small, when he starts to sit up on the floor, when he starts to walk, and even around the age of four or five. Some women explained the cause of this illness: 'Certain insects, *ing'ong'o*, bad insects, *idudu ibi*, very small, almost invisible, come into the house, they fly and fly, *likuguluka guluka*, and cause *ideje*.' Others told me: '*Ideje* is an illness which comes naturally. It comes like a dream, *lyomlota lilyo lideje kuko likuza:* it wanders around, then blows, *kumputa*, on a child, who becomes sick.'

Although different, the two versions refer to something ethereal, almost invisible, which moves in the air,[44] not actually the air but the *surroundings* of the baby, evidencing the ineluctability of this affection, feared as much as impossible to prevent, and for which there seems to be no efficacious cure. Should a healer manage to cure a child of *idejedeje*, he may ask for an ox in payment.

The illnesses that I have briefly described are the ones that most worry the mothers. There are also others of which the mothers are afraid: measles, *iselenyene*, chicken pox, *matetekuwanga*, cholera, *cipindupindu*, cardiac palpitation, *itunung'ha*, as well as repeated coughing, *ng'hololo*,[45] colds, *mafwa*, and stomach ache, *mchango* or *masangu* or *masungo*.[46]

One particular illness, of which the women do not willingly speak and of which they generally know very little, is *lawalawa*, which seems to affect the genitals of children at a very tender age. One of my informants told me that it is a relatively recent pathology, which appeared only 'in the time of Nyerere's government'. In females, the skin between the legs is flaky and the genitals take on a darkish colouration, *mugati*, 'they become black, *cikutukuwala ciwa cititu'*. In males, the penis, *nzunga*, more specifically the penis before circumcision, leaks drops of oily liquid, *mapotya (k)utongo-cispa*, while his lower abdomen is very hot. For both there is no medicine. While the male may be circumcised even though he is still breast feeding, the female requires an incision to be made to drain off the blood and this stops the skin flaking. After the operation, the genitals of both the males and females are washed, for two or three days, with water in which *mapande*[47] has been left to soak.

The non-illnesses

Alongside the true illnesses, mainly recognised as such because there is a change in body temperature together with other symptoms, mothers often find themselves faced with diarrhoea, sometimes accompanied by vomiting.[48] They use a series of distinctions in classifying diarrhoea in their children, and these reflect an elaboration that is strongly anchored to cultural models.

First of all, they distinguish the agents that promote diarrhoea.[49]

On starting mixed feeding, the food added to breast milk may cause diarrhoea, merely due to the fact that after cooking, it is often left in the saucepan for some days. Dust and other undesirable – and more or less visible – guests, are always lying in wait. When, furthermore, the child participates completely in the adults' diet, some types of food may be indigestible for his stomach, – beans for example, or some particular vegetables and an *ugali* which is too dense.

The women also indicate variations in temperature – night and day temperature range is significant – and other factors linked to seasonal variations, starting from the rainy season, as causes. As in the cited *nyawana*, they know for example that when certain trees germinate, *mpela* (baobab), *mnynga* (not classified) and *mpululu* (not classified), children may be subject to diarrhoea. The mothers are not able to explain the reason, but they have noticed the coincidence between the two phenomenon in small babies.[50] They also coincide with the baby's stages of development. In fact, the various stages in motor development, from sitting on the ground, to crawling, to the first uncertain steps, are often accompanied by diarrhoea.

Another important moment is the teething stage. The baby may have diarrhoea but, although it may be accompanied by a fever, the mothers consider that the child's health is not affected. The fact that the child continues to play, means that these symptoms are not indicators of a true illness.

In this way, although the appearance of diarrhoea does not cause excessive apprehension, the women do ask themselves what the cause may be, showing that the problem is never neglected or underestimated.

Traditional remedies are used at the very onset of diarrhoea. A mother tries to free her child from this fastidious problem, using her own knowledge and experience and everything that her mother has taught her and that has been tried out by other women before her.[51]

The baby is given the usual cream of millet and water, *uji*, to which, however, the fruit of the baobab, *upela*, ground to a flour or previously melted in water, has been added, for its astringent properties. A special type of millet, *muhoni*, is considered to have the same therapeutic qualities and is used when available. The paternal grandmother often goes looking for special roots, in the brushwood, *mbago*, which will be mixed in the *uji*, together with the *upela*.

If the mothers have correctly interpreted the cause of the diarrhoea (and of vomiting if present), they are sure that it will pass within one or two days, as they are not symptoms of a true illness, but merely normal inconveniences of growth or of seasonal variations. In fact, the child affected with this type of diarrhoea sucks normally, continues to do what he normally does and shows no physical change. In other words, the mothers say, he continues to be well and his development is neither altered nor slowed down.[52]

When, however, diarrhoea is refractory to domestic remedies, the mother's attention and worry gradually changes and intensifies, because of the persistence and frequency of the symptom. Asking herself once again about possible causes, her thoughts begin to go in other directions. The colour, consistency and odour of the faeces, the changes in the child's physical appearance and behaviour, which become more and more apparent as the days go by, open worrying scenarios for her.

To look for the cause means to understand its *nature* and, therefore, a possible remedy, in a crescendo of gravity that may put the child's life itself at risk.[53] The help of an expert therefore becomes immediately urgent. Sometimes, however, even this may not be sufficient: precious time may be wasted and appropriate remedies, if there are any, may not be used. This is the case in acute or chronic diarrhoea,[54] the causes of which are to be sought among human agents and about which the women never spoke of *mphungo*, true illnesses, but of the physical conditions of the child, therefore of a damaged or wasted child, *kuviza mwana*.

Among the human agents, bad or wicked people, *wasugusuge* or *wabalabala,* and witches, *wahawi* (sing. *mu-*), are decidedly among these, as, through jealousy or envy, they may cause early milk loss, or unexplainable alteration of the same in only one breast, consequently causing the baby to refuse the breast and to develop diarrhoea.[55] These two cases have a different effect on the fears expressed by the women when they refer to possible incidents to which their milk may be subject. The first, which has extremely dramatic consequences, is really rather rare, while the second, more common but more limited, may be resolved by the women avoiding breast feeding from the 'bad' breast and starting mixed feeding.

On the other hand, the fear of being stared at by a sterile woman is a different story, as it is seen as something more concrete, with immediate consequences:

> Someone may look at you badly with their eyes, *heya, munhu du yalanje vibi, kwa meso,* and, from that moment on, your milk will set off diarrhoea in the baby, so much so as to compromise his development.

these words made me understand just how much the women fear the proximity of certain women, when they are breast feeding or when their breasts are exposed.

Another case of deteriorated milk cited is when a person who has medicines, *ena sale,*[56] looks at a mother who is breast feeding, or passes behind her. The milk drunk by the baby will cause wounds on his head, *mapilime nu mutwe.*

The behaviour of the baby's mother herself may also be considered the cause of an alteration in the baby's health. From the moment in which I was able to overcome the women's reticence in allowing me to understand the specific maternal behaviour that, according to them, damages their children, new scenarios on the breast-feeding process were opened to me, scenarios which were unthinkable at the beginning of my research. I found myself, in fact, investigating the prohibitions placed on a woman's sexual activity, in her role as mother, nurturer and wife – prohibitions that are placed on men also, as husbands and fathers.

So it was, more than a year from the start of work in the field, that the moment had arrived to put hesitation aside and to face what I had been feeling for months – that which the journey with which I opened this chapter, begun in the dawn of a new and luminous day, had so clearly and directly brought me up against once again. I had, therefore, to study in depth those beliefs which maintain that the sexual activity of a mother, and of a father, has repercussions on the health of a newborn baby during the period of breast feeding. On considering the problem from another angle, I prepared myself to consider post-partum sexual abstinence, a behavioural model which is widespread, in different ways, in the whole of sub-Saharan Africa, and a theme widely dealt with in anthropological research.[57]

Post-partum taboos

Men and women's sexual behaviour is regulated by norms that the Wagogo learn and are called upon to respect from their adolescence[58] – norms that, as for all rules, foresee sanctions if they are not observed. Thus, with the birth of a new baby, the parents must respect sexual abstinence. While for the woman it is an obligation that she must follow for the whole breast-feeding period, for the man, it is limited to the three or four months following the baby's birth.[59] When this precept is violated, the sanction is particularly serious, in as much as it affects the transgressor through a third party.

In fact, the women attribute the origin of some serious forms of diarrhoea to the violation of this rule, and if these are not treated promptly, as tradition dictates, they cause a progressive and worrying alteration of the baby's health status, even leading to death.[60] How do the women formulate such a relationship between the violation of the rule regarding postpartum sexual abstinence and the deterioration of the child's health status?

In order to understand this correlation, it is necessary to return to what has been said about the concept of *mbeho*, 'ritual state'. This is because the social system as thought of by the Wagogo is the result of an equilibrium of forces, sensitive to the behaviour of individuals. Here, as elsewhere, the loss of 'equilibrium', the lack of 'harmony', produces a negative state, which reflects a symptomatic or ritual bad condition, *ibeho*. In order to preserve this social equilibrium, both men and women must scrupulously observe correct behaviour and prohibitions are the instruments guiding the conduct of every individual.

As well as the above-mentioned diverse time range applied to them, we shall see, in detail, the different effects on a mother and a father brought about by nonobservance of the rule of abstinence. I therefore consider it useful to pick up on Murdock's indication, limiting the expression 'postpartum taboos' specifically to the norm which forces the woman to interrupt her sexual activities during breast feeding (Murdock, 1967: 161).

The close connection that the woman believes to exist between her body and her milk, all that she has been taught about the dynamics of conception and the consequent physiological balances which bind her to her child, are aspects of her cognitive experience which imprint her behaviour and her most intimate feelings. On speaking about the quality of maternal milk, it clearly emerged just how much the women consider this fluid as a participant in the vicissitudes of their bodies, so much so as to be altered every time any event changes their natural internal equilibrium. So, as we have seen,[61] we have the fever, or the prolonged exposure of the breasts to the sun's rays, or a too-long interval between one feeding and another, considered to be incidents capable of deteriorating the milk through the heating of the body or of a part of it. This process is visible, thanks to the diarrhoea with which the child is affected while feeding on milk in one way or another deteriorated and, as the women stated, 'no longer good for him, no longer "his" milk'.

The mother can easily find a remedy for some of these inconveniences. When, however, her sexual behaviour is called into play, it becomes more complicated. Not only may the consequences for the baby's health become very serious, but also her very image of 'good mother' may receive serious blows.

Thus it is, that in consideration of above all the obligation to observe post-partum sexual abstinence and the consequences that would derive from a derogation of the same, that a mother rarely breast feeds another woman's child or gives her own child to another. In the Gogo customs, there are only a few categories of women with whom, for a mother, at least in theory, such an exchange could occur: her husband's sister, her brother's wife, both of whom are indicated by the name *wifi*,[62] (a term of address between them), her younger or elder sister, and sometimes her own mother or mother-in-law.

These women, tied to each other by blood or affinity ties, become possible reciprocal nurturers only when certain conditions are present: (a) they must have delivered in approximately the same period; (b) they must strictly observe post-partum abstinence. Whereas the first condition is easily controllable, the second requires reciprocal, personal and direct knowledge and respect between the two mothers, as such a personal and intimate individual behaviour is called into question. With regard to this and when specifying the quality of the relations which bind the two nurturers, the women talked about *mulilumbi*, great respect, and *mlumbi myago*, trusted friends, as requisites which are at the basis of a possible exchange of children.[63] A woman must be able to trust another completely with regard to her sexual behaviour, because the quality of her milk, and consequently the health of the infant, depends on it. It is not surprising, therefore, how uncommon this custom is in daily practice!

When post-partum abstinence is not respected, the women distinguish the type of violation, according to the possible partner with whom the woman has had sexual relations. To have sexual relations with one's husband, or to have sexual relations with another man or men other than one's husband, assumes different connotations and consequences for both the woman and the child.

How can effective violation of the rule be proved and how to distinguish, with facts, the two different types of violation, if the subjects are not caught in the act? It is the diarrhoea itself that acts as litmus paper, proving that the violation has taken place and *with whom*. A sudden diarrhoea, with a specific consistency, colour and odour of the faeces, certifies a pregnancy following intercourse between spouses; a similarly sudden diarrhoea but which is protracted and has equally peculiar characteristics, attests extramarital relations.

Kutuga muda: a new pregnancy[64]

A Gogo mother who becomes pregnant during the period dedicated to breast feeding (worse still if this occurs during the first year of the newborn baby's life), is severely judged by all the women – 'a very ugly story', *ilema ibi sana*: it is the most evident proof of the mother's lack of care towards her newborn baby. The reason is that as soon as she realises she is pregnant, the woman is forced to stop breast feeding, *koseleza*, thereby depriving her child of the nourishment which, more than anything else, guarantees his correct growth.[65] The situation created by the new pregnancy is therefore extremely delicate for the baby's health, as he is suddenly deprived of the sole nutriment that can make him grow strong and healthy.

A mother experiences this early pregnancy with a deep feeling of shame, *cevu*: she is ridiculed by the other women for not knowing how to deny herself to her husband and, at the same time, reprimanded by the older women. In the past (although it does still occur today), the latter would sit in assembly and bitterly reproach the woman who had so clearly violated the post-partum taboos. It is clear that the atmosphere around the woman becomes strained and she herself is extremely uneasy in a situation in which her image as a 'good mother' is compromised.

The women, however, know very well that tradition offers them some expedients, to be used to prevent untimely pregnancies. Traditional medicine, prepared by some of the older women, to be tied to the *kanga*, and the use of coitus interruptus are both expedients used to impede the consequences of a sexual behaviour not in conformity with the norms. Abortion, as a technique used to interrupt an undesired pregnancy, is very rare and is unanimously judged as a very serious act, such as to threaten the woman's survival and to cause serious risks for the person procuring the abortion.[66]

> To avoid a pregnancy 'out of time', the man must 'put the water outside',[67] even though sometimes things do not go the right way, something doesn't work as it should, and so you find yourself with another baby in your stomach!

This may happen, if you are *very unlucky*, the women emphasise, during the act of sexual intercourse indicated with the expression *kupagata mwana*,[68] (literally to support, to hold the baby in your lap), and with which, after months of complete sexual inactivity following the birth, a father has only one encounter (sexual) with his wife and the mother of the newborn baby and which frees him from the obligation of sexual abstinence.

From what has been said so far, it is clear that a new pregnancy delineates a complex picture for the mother and the newborn baby. I will now try to rebuild what I was able to put together on the problems about which the women were often reticent, if not elusive. First of all, I had to look for an answer to the prohibition to breast feeding in the presence of a new pregnancy. This direction would take me to the connection between the woman and her milk and, for both, with the nursing baby.

> Nothing [my informants told me] heats a woman's body more than sexual activity, changing her milk. How can she not think about the consequences of her behaviour? How can she forget how strong and delicate is the bond which unites her with her child?

   With these interrogatives ringing in my ears, more like exclamations, I wanted to summarise the sense of the answers given to me in reply to my question as to why a new pregnancy impedes breast feeding to the last born.[69] From these answers, I could start out on the route to a deeper understanding of the equilibrium, repeatedly underlined by the women, existing between the mother and the newborn. I could only give full sense to this equilibrium by going to the origin of the bond between a woman, a man and their future child, at the time of conception.

   The majority of the women were concise on this theme, leaving me very few opportunities for going any further: 'It is the work of God, *zinguvu* (strength, power) *zo Mulungu*. It is always God who puts together: he knows, we do not know.' It was a traditional birth attendant and a few other women who spoke to me about it. Thanks to them, I was able to put together the various passages of fecundation, even though they confirmed that, when conception is accomplished, it is always a complex thing, special, that *only* God knows.

   First of all, it is believed that a single day, *siku monga*, is sufficient to become pregnant, *nda ikwinjila*, literally to enter, to go into the stomach.[70] In the act of conceiving a new human being there are two types of seed at work together: the male and the female seed. When, in sexual intercourse, the former finds the latter (if there is no female seed the male seed accomplishes nothing), they amalgamate and 'the baby jumps out, *ziwawunganika, yazatokea mwana*'. This encounter gives rise to what one elderly woman defined as 'the fecundated egg', *cisanga*: a combination of the two different kinds of seeds. It is, in fact, from a *combination* and not from a *summation* of the two seeds, that a new individual begins and constitutes the special and unique bond which unites the parents to their own child – a special and equally peculiar bond having, in its turn, its own special equilibrium, different for each child, in as much as the possible combinations are practically infinite. Furthermore, from this same synthesis, a series of similarities arise, now with the father, now with the mother, as also with the ancestors.

   The women did not fail to underline that, if the child being formed receives his bone structure from the father's seed and his soft parts from the mother's, both of the parents give the child their own blood.[71] So the baby's blood is a mixture and, although belonging to both the mother and the father, the result is something unique, new when compared to both: 'Each individual has his own blood, *munhu na sakami yakwe*.[72] Every blood has its own direction, *misipa*[73] *yakwe*, literally his veins.'

   It is from this encounter and recomposition that the intimate bond is created, the equilibrium which is so important as, in certain circumstances, the life of the most vulnerable of three individuals, the newborn, depends on it. It is never to be changed, above all with different blood, for the baby would run grave risks. It is for this reason that if a woman has intercourse during pregnancy with men other than her husband, she risks abortion.

Conception marks only the beginning of the reproductive process, which to be completed must proceed for nine to ten moons, allowing the baby's complete development, and to which, as we have seen, the father contributes for a few months.

In this way, when the sexual act has given rise to the encounter of the two seeds, the vagina closes, *lyadinda*, it is sewn, *lyahona*, the women told me metaphorically, thereby not allowing out blood anymore. The blood will come out once again when the baby is born; now its task is to build the baby, *yoyikuzenga mwana*. In fact, once conception has taken place, blood in the vagina coagulates, forming three *vidonje*, small clots which arrange themselves like the three stones of the hearth, and from which the whole takes its name, *mafigwa*. There, right in the centre, the fecundated egg places itself so that it may progressively be enveloped by the three small spheres that will slowly develop into a container, in the cavity, in which the baby will develop.[74] This container grows with the baby until it becomes a thin membrane, *ikuli*, literally a thin layer of skin, a film, which when it contains the fully formed baby is called *nguwo*, the term with which the women intend to indicate the 'house in which the baby is', *wakutya yo kaya yono yali yekaliye mwana*, in other words, the placenta:

> *Nguwo* [my informants told me] is formed of the mother's blood which previously, coagulating, *iwakiyenda*, in the three *vidonje* has welcomed the fecundated egg, *cisanga*. *Nguwo*, is now needed to tie the baby, *yademuka mwana*, stopping him from going away, *yawuka*.

When, after the baby's birth, the placenta comes out it assumes the name *igonga*, a typical term of the Gogo tradition, rarely used today and substituted with *linyina za munhu*, meaning the baby's mother – just as the mother looks after the newborn baby, the placenta holds the baby and looks after it from inside.

The full value of the bond uniting parents to their offspring appears from this description, a unique bond in its formation which, once constituted, must be maintained in its correct initial equilibrium, that is, without being compromised by external events or persons. It is, par excellence, his *mbeho swanu*!

Among the events that can alter this normal equilibrium, damaging the invisible thread which unites the infant to his mother, is a new pregnancy, the first effect of which is to increase the woman's body temperature and to make her milk become colostrum once again. When the baby sucks this milk, now 'the milk of the baby not yet born', he starts having diarrhoea and vomiting.[75] The consistency of the diarrhoea, liquid and solid together, its colour and smell as mother's milk (the vomit itself looks like milk which has gone off) – all this indicates a new pregnancy.

While these are the most evident signs of a new pregnancy, a mother has only vague ideas about the beginning of the same. If a missed menstruation, *siyakwitumila*, is the first sign of a possible pregnancy, it is true that in general, thanks to breast feeding on demand, a prolonged amenorrhea leaves the woman without this precious indicator of a new gestation. The women simply say that they *feel, kulamula* (literally to decide), when they are expecting again. It may happen, in fact, that in the absence of diarrhoea and vomiting, a mother may continue to breast feed normally in the first months of pregnancy – as I was able to see in the village, where there were children who had a small sibling after eight months after weaning, and others after only four to five months.

In any case, on being faced with such *eloquent* diarrhoea accompanied by vomiting, it is very important for the mother to realise rapidly that she is pregnant and, without hesitation, to act consequently. As well as stopping breast feeding, she must obtain a special remedy able to free the child from her bad milk which he has swallowed, thereby putting an end to the diarrhoea and vomiting.

In the past, the woman went to the traditional healer, who prepared a special antidote by cooking the fat of a sheep's tail, *mafuta ge ngh'olo* – believed to have a cooling effect on the baby's stomach, heated by the mother's milk – with a substance extracted from a parasite plant. The compound thus obtained is cooled and then slowly drained into the baby's nose, with the help of a piece of animal skin shaped into a funnel. In general, today, the mother goes to a healer or she herself prepares the sheep fat, frying it in tiny pieces. She then adds it to a paste made from wholemeal millet flour, *ngh'ata*. In any case, the women say that there are no longer any experts capable of introducing the medicine through the nostrils, as was done in the past.

After having taken the remedy, the baby is lain or seated in the sun and in a short time the 'bad milk', present in his stomach and cause of the diarrhoea, will be expelled thanks to the medicine. Its task is in fact to provoke further diarrhoea, a *curative* diarrhoea, which will clean the baby's stomach of the source of heat represented by the 'bad milk' swallowed.[76] From this moment onwards, the baby must be looked after with particular attention, as he can no longer count on maternal milk to grow healthy and strong. One woman's words describe well the condition of a pregnant mother with a small child to raise:

> A woman is only one person and she has to carry two weights: one in her stomach and one on her back. A truly onerous and shameful condition, a situation to be avoided. Only one baby at a time should be raised.

These words convey all the workload involved for a woman who becomes pregnant too early. Not only does she have to take care of her pregnancy, but she must also pay special attention to the weaned baby, multiplying her commitment and her efforts in terms of feeding and loving care, if she wants her child to grow without dangerous delays in his development. The same care dedicated to the quality of food must also be shown in terms of the quantity that he must eat during the day. Both quality and quantity may become problems in an environment that depends so strongly on the rains and where the variety of sustenance considered suitable for a small child's diet is extremely poor and with scarce organoleptic properties. All of this, it must be remembered, is added to the woman's multiple, daily work commitments.

It is therefore not surprising, considering the environmental difficulties on top of the women's workload, that little spacing between successive pregnancies exposes the child, weaned so precociously, to an inadequate diet, with protein-caloric consequences,[77] and perhaps some lack of affection, not easily attributable purely to maternal negligence.

Mhangahango: a mixture of sexual relations

If violating the post-partum taboos demonstrates the little care a mother has for the well-being of her baby – 'a mother plays with her own child's life', as the women declared in the case of a precocious pregnancy – to have sexual relations with men other than the husband signifies killing one's own child, *kukopola.*

In tracing the behavioural lines of a breast-feeding mother, all the women clearly highlighted the difference between breaking the rule with one's own husband, to whom the paternity of the wife's[78] children is always attributed, and breaking them with a lover. Moreover, in sexual activity with men other than one's husband, the women never failed to differentiate between a lover present from the time preceding the pregnancy, (therefore a more or less constant presence in the woman's sexual life), and an occasional lover. The former has many of the obligations of a husband; for example, he must observe the same sexual abstinence that a father must observe with the baby's birth:

> If you always have the same lover from the time of pregnancy, you may continue to see him after delivery, but as a friend, *mbuya ya mlomo.*[79] You may even sleep with him, but without having intercourse, *hasina mtambo.* You must be able to trust the man completely, *mwaminifu hawaha.* If that is not so, the baby is killed, *siyo mwaminifu yakukuwulajila mwana.*

In order to understand more fully the distinction between lovers, as made by the women during the time of my research in the field, it is advisable to make a brief digression into what Rigby wrote about 'institutionalised adultery' – the relationship between a married or nonmarried man, *mbuya,* and a married woman (1963; 1969b).[80] The relationship defined as *kwilajila,* literally to show oneself,[81] was negotiated between the two parties, – the husband, an old man, the young wife and her young lover – it remained secret, at least in theory, and implied no modification of the status of the actors involved. In a stable relationship, it was possible for this bond to give way to limited cooperation in agricultural and pastoral activities. The agreement could end at any time, with one of the parties breaking it. A variation of this theme was represented by the exchange of the right of access to the sexuality, *kusutana,* of the respective wives.

As the women were my informants in this case, the husbands became the subject of the exchange, *cisutane na mlume*. On the other hand, Rigby himself pointed out that it is frequently the wives who start off this kind of exchange. A woman may provoke the wife of her lover into becoming the lover of her own husband, as long as the men are of the same age. Such a relationship is allowed only between two married couples: 'Two couples are companions, three are a crowd', my interlocutors told me.

Bearing in mind the picture portrayed by Rigby, the use of the term *mzelelo*, lover, which the women used when talking to me, alternating it with *mbuya*, quickly suggested two aspects for consideration. The first was that, using the term *mzelelo* (literally young man), they were frequently speaking of a casual partner, who had little to do with the *mbuya*, a much more structured figure in the Gogo tradition. The second was that, thirty years on from Rigby's research, not all sentiments felt towards the canons of sexual morality had waned, as the use of two terms for lover indicated the (still) two distinct types of relationship. In the light of these considerations, care must be taken not to make the mistake of using the two terms interchangeably, as synonyms.

These more or less veiled conversations on themes regarding sexuality made me understand, between jokes and derision, just how much sexual activity is a subject where the two genders, almost always separately, gave rise to chattering loaded with witticisms and malice. When, however, the subject of the conversation touched on the sexual activity of a mother during breast feeding, the interlocutors' attitude changed and became more serious, according to the degree of superficiality and thoughtlessness attributed to a woman's behaviour.

Extramarital relations by a mother are negatively criticised by the community. It is proof that the woman is a 'bad mother': she is inconsiderate, doing something very bad, *kukodya*, to her child:

Agwe walya, ivunhi yagana yunji, you have eaten until you are full and some-one else is feeling bad.

or

Walya masina, waleka wuci, you have eaten the cells (of the beehive) and now you have no more honey.

These were metaphors used by the women to indicate risky and impru-dent behaviour: eating too much or eating the cells leads to serious con-sequences. In the same way, the expression *uzidi cipeyo*, to break the rules (literally meaning to fill a container to more than its capacity until it spills over), or *umzumphile itembe*, to have intercourse with different men (literally meaning to jump the roof) signify the damage that a mother causes her child by having too many sexual relations and by mixing dif-ferent men.

As in the case of a new pregnancy, women blame the mother for her body's altered equilibrium. They explain it as being the result of her having had sexual intercourse with a man other than her husband:

> The woman's and the man's blood refuse each other when they meet during sexual intercourse – *alu isakami iwabita, kilema*. The woman's body overheats and this spoils the milk, causing damage to the baby as soon as it sucks it. This may not happen, but only if the woman is lucky.

Mixing different men, *hono nhanzahanze na wakunze*, produces – as my interlocutors told me – a completely new disorder, a mixture, *mhanga-hango*, totally unknown to the woman's body. From here, the harmonious equilibrium created between the father, the mother and the new human being at the moment of conception is interrupted, transforming the mother's milk into 'bad milk' which damages the baby, *komtima mwana*, causing him a serious form of diarrhoea.

As in the attempt to avoid a new pregnancy, the women pointed out that some of them try to escape from the consequences of their actions, by preven(ta)tive resorting to special medicines, *miti ye cigogo*, which should avoid any damage to the little one. I found out that there are methods which can be used on the child, on the mother or on both – washing the baby with special roots left to soak in water, or spreading medicine around his waist or on the mother's breast – used with the precise aim of avoiding all negative consequences. The women make use of these methods secretly, although, as they normally need money to pay for the medicines, they ask their husbands for this, making up excuses for going to the *mganga* to treat some kind of unexplained malady.

If the mother knows that she has had extramarital relations and the baby does have diarrhoea, it is its resistance to domestic remedies and its characteristics, the women say, which should alert the mother as to the cause of the changes taking place in the physical state of the child. It starts with a watery diarrhoea which, after a few days, becomes bloody and milky with a particularly unpleasant smell and is often accompanied by vomiting. At the same time, the baby loses appetite, sucks with difficulty, becomes more and more listless, *yalimbeho*, lacks energy, *yalabalayelabal-aye*, becomes weak, *mdhaifu*, shaky, *akucicimacicima*, visibly regressing on the motor level as the days go by. In fact, the women underline with a certain emphasis that, if the baby has learnt to sit up or crawl, stand up or walk, he rapidly regresses, not having the strength to do anything. They told me:

> The baby loses weight, he has a low body temperature. As the days go by he begins to look like an old man with wrinkled skin, his hair becomes fine and thins out, taking on a reddish colour. He is so thin that his tail comes out and when he tries to walk, his gait is similar to the stirring of twigs, *cikuneng'a-neng'a*, in the slightest breeze.

Before reaching such a situation, the mother should feel it necessary, if she knows that she has broken the rules, to immediately resort to a specialist who knows the appropriate medicines, as tradition dictates. The specialist is a woman, *mdala*, an elderly woman.

The therapy is a treatment in which rite and cure are combined. The woman and her lover secretly go to the healer who, after having listened to them and having understood that the child is sick because of the mother's adultery, looks for the correct medicine, *mapande*, for that specific case, *msace mapande, ukutumya miti ye cigogo, cigogo du ye cinyegi*. Even the way of collecting the piece of bark must adhere to precise rules: the bark must be fresh, cut from a certain tree, in a certain way and must fall to the ground in such a way that the internal part is facing upwards. The *mapande*, collected in this way, is then placed on the threshold of the healer's hut and the woman and the man must jump over it three times together. Only then may the child be treated. The piece of bark is left to soak in water in a container in which the baby will be washed four times, in the two days following the visit to the *mdala*. Once the therapy has been completed, the woman must throw away the water, the medicine and the container in a place indicated by the healer. The latter will have indicated also the tree under which the used objects must be buried, at dawn on the third day.

I found some differences between what the women told me about the therapy. Some were insignificant, whereas one was particularly significant in that the couple were to have intercourse followed by a bath, – to cool their bodies[82] – before starting the baby's treatment. Apart from the possible variations, the good result of the cure depends on the early diagnosis and rapid treatment.[83]

From what has been said so far, it may seem that an extramarital relationship during breast feeding may lead to consequences which are less serious for the mother and her child than a new and precocious pregnancy. By dealing with the crisis immediately, the child can be returned to good health and the mother can continue to breast feed him, which she cannot do if she is pregnant. In reality, things do not go according to the description: theory and practice are very far apart. The women told me:

> A mother goes from door to door with her baby, with no success. She goes backwards and forwards to the hospital of Dodoma, but good food and medicine is not enough. The baby is not well, *mwagonga*, he still has diarrhoea, *pwa-pwa*,[84] and, in the end, he dies. She should have sought out the *mapande*.

With these words, they wanted to underline the inefficacy of both 'normal' traditional therapies, going from door to door, that is from one *mganga* to another without results, and Western medicine, – good food and medicine do not heal the child.[85]

What stops a mother from going in the right direction? What stops a mother from having the 'right remedy', able to save her own child? Before looking for the answers to these questions, I would like to answer another question: what happens when it is the baby's father who has broken the rules?

In this case, the situation is even more dramatic, as it involves the sexual behaviour of a man, husband and father, a terrain to which only a few women have direct access.

We know that when a man becomes a father, he must observe certain obligations and, among these, the ceremony denominated *kupagata mwana*, literally holding the baby, is extremely important. It consists in his having one single act of sexual intercourse with his wife after three to four months of complete sexual abstinence after the baby's birth. This single act frees him from abstinence, allowing him from then onwards to pick up his normal sexual activity once again.

When, however, he breaks the rules, *utambuce mwiko*, (literally *kutambuce* means to trip up) the consequences for the baby's health are the same as those brought about by the mother, *komtima mwana*. It is enough that a father comes into contact with his baby, directly – for example, taking him into his arms – and indirectly, – simply by stepping over the piece of material, *tambuka sambo ya mwana*, that the mother uses to hold the baby during the day, for the damage to be done.

It is thus that a mother, faced with the worsening conditions of her child's health in the very first months of his life and after having tried various cures with no success, will start to ask herself with increasing anxiety about the reasons for her child's condition. If she is certain she has not gone against any taboos, she will begin to fear the involvement of the baby's father, her husband, and she cannot speak to him about his probable bad conduct. The only persons with the power to investigate, in this direction, are his mother and his grandmother.

It is understandable that this will mark the beginning of tension, conflict and anguish inside the *kaya*: first of all, for the insistence and the embarrassment with which a mother seeks help and action from her mother-in-law; and, secondly, for the father's resistance to admitting, if this is the case, his responsibility for the poor health status of his little one.

What is certain is that these conflicts delay action and can be – and often are – extremely dangerous for the baby's life. Diarrhoea that is not treated promptly with the replacement of lost fluids rapidly causes a state of dehydration which, together with other possible pathologies that are often present in these cases, can seriously compromise the baby's health, ultimately leading to his death.

Notes

1. My reference to the binomial wife-mother is dictated by the fact that it was, at the time of my research, a widespread model in the village. While the figure of the single mother remains in the category of exceptions, pregnancies in widows or divorcees are not infrequent.
2. Among the Wagogo, a woman's sterility, like a man's impotence, means that the marriage may be dissolved. In both cases, however, the decision is made after careful thought, confirming the great emphasis that the Wagogo place

on affinity ties (Rigby, 1962b; 1966c; 1967a; 1968a; 1969b). In fact, for a man it may only be a temporary state, especially if he has other wives and children, while for a woman, a specialist is consulted, *mganga*, and he will give her medicine to have children. The women told me '*Nghelecelaje mbeleko*', the *mganga* gives the medicine for procreation. *Kuteleka*, literally to cook, when used with the medicine, *kuteleka miti*, assumes the meaning of receiving the medicine, the treatment from the healer, and *mbeleko* indicates the power of reproduction, of proliferation, of bearing fruit, basically, of fertility. On the elaboration to which reproduction and fertility are subject, see Loizos and Heady (1999), Moore, et al. (1999), Tremayne (2001).

3. There are many countries in sub-Saharan Africa in which reaching the status of adult woman depends on the young woman's ability to reproduce and often, to reproduce a male (Caldwell et al., 1992; De Bruyn, 1992; WHO, 1994; Ankomath, 1998).

4. Explicit reference to what the young mothers learn at the village clinic, where the use of a calendar is a sign of modernisation.

5. The women are subject to a few prohibitions with regard to food to safeguard their maternal function. For example, a forbidden food for the fertile woman is cows' udders, *imisi*, because they would stop the woman having milk. 'The baby sucks and the milk doesn't come', the women told me.

6. The same attitude is present in other groups, for example the Fulani, where the women avoid any reference to their state, so as to protect themselves and the child they are carrying (Johnson, 2000).

7. The women used to alternate indifferently *malenga*, water, or blood, *sakami*, referring to sperm, in this specific context.

8. Babies born 'dirty' are those covered with a caseous film, or with tinted amniotic fluid at birth, attributed by Western medicine to premature births or to intrauterine sufferance. The Wagogo interpret such cases as proof of violation of the rules on sexual behaviour; thus the newborn baby's body, *mchafu*, is the proof of outrageous behaviour.

9. We will see this difficulty on dealing with post-partum taboos, from which the different expectations placed on a mother and a father will emerge.

10. As well as Raphael (1966; 1981), see Kitzinger (1978); McCormack (1988); Raphael and Davis (1985); Glen et al. (1994); Rich (1995).

11. I have already spoken of 'bad milk', as a result of changes occurring in the mother's body (Chapter 3). We will shortly see how sexual activity can alter the harmonious, if fragile, bond between a mother and her baby.

12. Kitzinger explicitly speaks of 'a kind of extension of pregnancy' (1994: 68) to underline the close relationship between the mother and her newborn.

13. Awareness of the dangers of the very first days of the life of a child, when he still has a piece of umbilical cord attached, is also found in other groups, for example, the Beng of the Ivory Coast (Gottlieb, 2000) and the Fulani of Guinea Bissau (Johnson, 2000).

14. The word *moto* literally means both 'hot' and 'fire'. On talking about the quality of milk, we have seen that hot milk is 'bad milk', and we will see that the same negative meaning is applied to the baby's body temperature: a 'hot body' or 'a heated body' is, for some reason, a body which is not in good condition and which is, therefore, exposed to danger.

15. This is true for how one is born and for how one grows. On talking about purification rites, Rigby also examines unusual events relative to birth (breech births and twins) and some characteristics of the newborn baby's develop-

ment (growth of the upper teeth before the lower), as events which may alter the well-being of an area, *mbeho swanu*, with the consequent need to recapture the 'normal' state through the performance of special rites (1968a). Birth and growth are surrounded with beliefs that no group is exempt from. Among the Igbo, for example, the coming-through of the top teeth is considered extremely abnormal (Agbasiere, 2000: 73).

16. *Hewa*, Kiswahili word, means 'air'.
17. *Mbeho*, literally wind, also means 'ritual state', good, as the women say above, or bad, *ibeho*, literally a strong or violent wind.
18. The fall of the umbilical cord also coincides with giving the baby a name. For the Wagogo, the baby is a person from birth itself.
19. Here the adjective *-chafu*, dirty, means 'something wrong', 'something not going the way it should' in the baby's body. When it is used for a baby born dirty, as we have seen, it indicates both something which is directly observable, the oily substance which covers the baby's body, and something which is indirectly certifiable, disrespect of the rule which limits sexual activity in a pregnant woman.
20. *Yecewela*, from the verb *kicewela*, specifically indicates the state of illness which affects a newborn baby for the first time.
21. This sense of bad fortune, as something unavoidable, is the same expressed by the women when faced with bad-quality milk, *mele mapelupelu*, which a mother experiences from the start of breast feeding and for which there is no easy remedy. The opposite, good fortune, *lweru*.
22. We find here the same knowledge of illness noticed by Janzen among the Bakongo (Democratic Republic of Congo, formerly Zaire) who recognise two causes of illness: 'illnesses of God' considered natural, and 'illnesses of man', the causes of which are to be found in humans or in supernatural forces (Janzen, 1978).
23. Another expression used for the same concept is *sakami hono ivine, ukutamwa* – if the blood dances, you will become ill, where the verb to dance, *kuvina*, always indicates a situation of agreement, of consensus, between one's own blood and the illness, negative for the individual in this case.
24. At the time of this research, national infant mortality was 104 per 1000, (World Development Report, 1990).
25. A 'hot mouth' is unanimously recognised as a sign of alteration of body temperature, to which women pay a great deal of attention, as they deem it an indicator of the onset of a more or less serious and more or less transitory sickness.
26. We have already found these adjectives on talking about milk (Chapter 3, note 29).
27. Other expressions used by the women to describe their children's progress are: *itombo ipelu* and *itombo ikamu*, which indicate, respectively, a baby who walks precociously and one that starts walking late. *Itombo* is the breast, the adjective *-pelu* means light, simple, easy and, for extension, something easy to do, its opposite *-kamu*, hard, consistent, difficult and, by extension, something difficult to do.
28. The verb *kumwalabata* and its synonym *kwembelwa*, used by the woman in her articulate response, indicate exactly the passage between a state of health and one of illness. They correspond, approximately, to the Italian expression 'covare una malattia', 'to hatch an illness'. (Literally: to pass the hand over the body of a baby to get the feeling whether his body temperature is high or normal).

29. There is a wealth of literature on this subject. I mention here Ayodele Ade-
    tunji (1991); Iyun (1992, 2000); Gottlieb (2000); Johnson (2000).
30. The case of the woman in the opening of this chapter goes against this rule.
    Her exceptional behaviour went hand in hand with the gravity and complex-
    ity of the case on a clinical, as well as cultural, social and psychological level,
    as we will understand further on.
31. I sometimes saw the clinic staff resort to the *balozi* who, in the absence of a
    husband, far from the village for a number of days with the herd or away for
    some other reason, may give his consent to transfer the child to the hospital
    in Dodoma. In the cases I observed, the children were seriously malnour-
    ished, with the symptoms of *kwashiorkor* or marasma, or both.
32. I remember that when they mentioned these, linking them to possible
    changes in the quality of milk, they almost seemed to be regretful or embar-
    rassed about having allowed me to participate in such a peculiar aspect of
    breast feeding, so much so that they became reticent. I was to understand
    later the reason for this reticence. As I had learnt not to do early on, I never
    insisted in this kind of situation, trusting in the advent of better moments
    when, above all, with time and more opportunities for dialogue, I would
    become a more significant (*visible*) person for them.
33. In this distinction, it is possible to find the general picture proposed by Janzen
    with regard to the populations of Bantu language of sub-Saharan Africa
    (Janzen, 1989).
34. *Homa*, fever, is a Swahili term, often used by the women, in our conversa-
    tions. The term is also used to cover a range of diseases, if one does not wish
    to say exactly what kind of disease he has.
35. *Indosi* or *idosi* really indicates the membranous zone, not yet calcified, of the
    baby's cranium. When the fontanelle closes, that part of the head is called
    twalilo, indicating exactly the part of the head on which the women carry
    water gourds or bundles of firewood.
36. The prefix *ci-* indicates something small, in this case it indicates the body of a
    few-months-old baby.
37. The greenish substance placed on the fontanelle is obtained by soaking par-
    ticular leaves in water, then chopping them until they have a creamy consis-
    tency. According to circumstance, they may be remedies prepared by the
    mother, by the older women of the *kaya*, generally the baby's paternal grand-
    mother, or by a specialist, and on the father's request. The quality of the rem-
    edy depends on the particular case.
38. *Kutowa cisangu* corresponds to Kiswahili *kupiga bao*, literally to beat the board,
    or to divine with the use of the divining board. Among the Wagogo, other
    objects may also be interrogated: shoes, *vilatu*, pieces of wood, *vipiji*, small
    shells, *viciliamba*, which when thrown on the sand are able to speak to the
    mganga on the basis of the position they have assumed. 'You can see what has
    happened', one *mganga* told me. The term *mganga* indicated by Rigby as
    specifically referring to the diviner, a different person from the healer, was
    used, in my experience, by the women in referring to both. In the case nar-
    rated to me, the person who makes the diagnosis and the one who gives the
    medicine, *kutowa miti*, clearly seem to be one and the same person. Apart
    from the terminology, the distinction between diviner and healer remains
    among the Gogo women.
39. *Kugombana*, verb of the Swahili language, indicates an exchange of strong
    words, to disagree, to argue.

40. The term *laho*, as Rigby noted (1969b: 168) is a term adopted by the Sandawe, *la'o*, a non-Bantu idiom, spoken by northern neighbours. *La'o* means literally the act of destruction, the act which produces disharmony, and it is used in this sense by the Wagogo, on referring to the actions of witch doctors or wicked people, or to disputes which put relationships between kin or relatives at risk. They told me that '*laho* is like threats made to a person after the use of strong words'.

41. According to one healer, the baby affected by *nyawana* has no water and the blood in his body is finished, *akosa malenga, yamala sakami mumuwili*

42. Some women confirmed that there is another herb, *mapangalale*, which should be boiled in the water to be used for washing the baby.

43. From the women's description, above all with regard to the convulsions without fever, it would appear that we are faced with epilepsy, even though the term used specifically indicates convulsions in a child. The term *idejedeje* is not used for an adult with convulsions.

44. A *mganga* also indicated convulsions using the Swahili term *degedege*. From the lexical point of view, this term recalls 'something which flies' in the same way as the Gogo term does. *Dege* is also a kind of butterfly, an animal, therefore, which moves in the air. *Ndege* is a general term for bird.

45. When the cough is persistent and resistant to all treatment, the women think that it is caused by the unusual size of the uvula, which must be reduced.

46. Stomach ache presents with different symptoms and causes, apart from the pain expressed by the child's crying. It may be linked to food, or due to worms, as the Swahili term *mchango* indicates. Hot water and sugar seems to be the treatment adopted by the women, but in the most painful cases, resistant to all domestic treatment (as for example, when the mother notices greenish streaks appearing on the child's stomach), hot water is not enough and medicines from the *mganga* must be sought.

47. *Mapande*, literally bark. The term indicates peculiar medicines obtained from the bark of some special trees.

48. In the village, as in general in the whole of Tanzania, the incidence of diarrhoea for the survival of children is dramatic. In their study on malnutrition in Tanzania, Kavishe and Yambi (1991) found that diarrhoea is common in children between 6 and 24 months, while Leach (1990) and Mwaluko et al. (1991) indicate diarrhoea as the main cause of morbidity and mortality in children under five years of age. This situation is found in all Third World countries (Bern et al. 1992; Davidson and Meyers 1992). In UNICEF's State of the World's Children (2002), diarrhoea remains one of the main causes of mortality for children under the age of five.

49. I dealt with the theme of diarrhoea in a previous article (Mabilia, 2000), in which I spoke about 'precipitative' agents (p. 195).

50. Studies have shown how climatic variations, together with seasonal cycles, may lead to critical periods for the health of infants due to the presence of pathogenous agents (Barrel and Rowland, 1979; Mutanda, 1980).

51. While, in the past, water was considered dangerous during bouts of diarrhoea, because it would increase the number of evacuations, today the women have learnt from the clinic that water should be given to a child with diarrhoea. In the clinic, they have assisted at the preparation of the drink to be given to the child. Heated water in which small quantities of soda, or salt, and sugar, *vicete, vimino, visukali*, have been added, is sufficient to prevent dehydration. While the women demonstrated, in words, what they had

learnt at the clinic, I rarely saw this teaching put into practice, for two reasons: firstly, due to the unavailability of the necessary substances; and secondly, due to resistance to change which, by questioning consolidated beliefs, is not easily accepted.

52. In reality, there are many illnesses which present with diarrhoea not considered as a symptom of a particular pathology. What Gogo women think about the different kinds of diarrhoea, not necessarily associated with bad health, is common to many African knowledge systems, from which diverse attitudes, behaviour and remedies derive (Ahlberg, 1979; Yoder, 1981, 1995; De Zoysa et al., 1984; Green, 1985; Kenya et al., 1990; Hogel et al., 1991; Green et al., 1994).

53. In the study conducted by Green et al. (1994) in Manica Province in Mozambique, and in that of De Sousa (1991) conducted in the south of the same country, on local conceptions of diarrhoea and the illnesses connected to the same, we can find the same attitude as that of the Gogo women on considering the causes of diarrhoea according to a progression of increasing gravity.

54. In referring to different types of diarrhoea I have used the distinction between acute and chronic or protracted made by De Giacomo and Maggiore (1999).

55. See Chapter 3. On thinking about a wicked act, an act caused by witchcraft, *uhawi*, the woman may be commiserated for her bad luck, but, sometimes, she may also be regarded with suspicion for having, in some way, brought envy or anger on herself.

56. *Ena sale*: this expression refers to subcutaneous cuts, generally made on parts of the body not exposed to indiscreet eyes, and on which medicinal substances, are placed or spread. These substances are made by witchdoctors or witches on request of people who wish to damage a particular person. The same procedure may be used for 'preventive cuts', defending the person from possible attacks from wicked people.

57. There are just a few studies: Saucier (1972); Bleek (1976); Caldwell (1985); Caldwell and Caldwell (1977, 1981); Orobuloye (1979); Lesthaeghe et al. (1981); Ojofeitimi 1981; Schoenmaeckers et al. (1981); Aborampah (1985); Isenalumhe and Oviawe (1986); Oni (1987); Hassig et al. (1991); Dean (1994).

58. Anthropological studies, starting from the classical works of Kenyatta (1938) and Richards (1956), have widely testified as to just how much African societies had organisational principles and moral values aimed at regulating individual behaviour and community dynamics. Among these, sexual behaviour was made the subject of an educational journey, in which rules and prohibitions had to mark and address women's and men's behaviour. Initiation and puberty rites, as in the case of the Wagogo, represented and represent, even though not so strongly today as in the past, the formal environment in which traditional teachings are transmitted to new generations. The changes that historical and anthropological research have revealed through the years have shown how much the so-called modernisation process, introduced firstly by colonial states, and then, by the modern African nations, has paid little attention to the contents of these extremely complex and totally new dictates for Western culture, misunderstanding them, due to ignorance or, worse still, to contempt, considering them primitive or even amoral (Beidelman, 1982; Ahlberg, 1991).

59. While Moller (1961) reported exactly four months, the women never spoke of a fixed interval (Mabilia, 1996a: 204).

60. The idea of a close connection between bad behaviour and the well-being of the individual, but also of groups and of the whole community, is widespread among many African traditions, but not only African ones, as witnessed by the work of Brandt and Rozin (1997). The association between diarrhoea and the bad behaviour of one or both parents has been witnessed, among others, by the studies of Cutting et al. (1981); De Zoysa et al. (1984); Green (1985); Janzen (1987, 1989); Last (1993); Green et al. (1994).

61. See Chapter 3.

62. The respective husbands address each other with the name *mbuyane,* while the term which defines their relationship is *mulamu wangu,* my brother-in-law (sibling-in-law).

63. For the friendship and respect, the basis on which the decision to breast-feed or to give a child to another woman to breast-feed is founded, co-wives are excluded from this possible exchange. Having the same husband implies, not rarely, being in competition for his wealth, his herds and in earning his attentions. Even having children, and healthy children, represents wealth for a woman which may promote envy among a man's wives (Rigby, 1962b). All of these are situations which do not promote that reciprocal trust which is indispensable to allow two women to exchange their infants in breast feeding.

64. In dealing with problems so closely bound to intimate aspects of the lives and sexual relations of men and women, I must state beforehand that my research was oriented above all by the meetings I had with 37 of the 114 women of the sample. In fact, thanks to them, the traditions, beliefs, rules and prohibitions, and their interactions with personal problems and pyschological impediments, gradually led to issues in the sexual sphere, notoriously difficult to deal with, and which gradually gave body to what, until then, I had only sensed. The information collected, thanks to this small restricted group, then became the subject of structured interviews with the whole sample. Together with them, I closely followed up, on a daily basis, seven cases of children under three years, affected with acute or chronic diarrhoea.

65. As many studies have revealed, the obligation of abstinence during breast feeding aims at ensuring the correct development of the newborn, while my interviewees rarely, and always marginally, mentioned the question of the mother's health, Furthermore, it is well to remember that prolonged sexual abstinence is in no way related to birth control (Bleek, 1976; Okediji et al., 1976; Caldwell and Caldwell, 1977, 1981; Dow, 1977; Orubuloye, 1979; Page and Lestaeghe, 1981; Schoenmaekers et al., 1981; Oni 1987; Dean, 1994). But, in reality, subsequent pregnancies further weaken the women's bodies, already tried by malnutrition and anaemia (Ojofeitimi, 1981; McCormack, 1988; Isenalumhe and Oviawe, 1986; Hassig et al., 1991; Maher, 1992).

66. Some women told me that the woman procuring an abortion could even be put to death, if identified. With regard to the birth control methods offered by Western medicine, and of which the women from the village had heard about in the clinic or the hospital, they were afraid of them, fearing that they could impede new pregnancies when desired. *Majira,* a Swahili term (literally time, period, season) is also used for contraceptive, often misspelled as *majila,* due to the little use of the letter *r* in the Gogo alphabet (Cordell, 1935: 5) – *majira or majila* because, in the same way as the seasons alternate, it is a periodical and not continual method, *nzilo.*

67. With the expression 'to put the water outside', which stands for coitus interruptus, the women always indicate sperm here with the term *malenga,* water.

68. In this custom it is possible find some similarities between that of 'crossing the poles' of the Kgatla (Schapera, 1971) and the rite of *ukupoko mwana* among the Bemba (Richards, 1939).

69. This belief is widespread well over the African borders, including the Western world, with its own modalities of time and behaviour. See, for example, the habit of giving the newborn to wet nurses (Fildes, 1997).

70. The expression 'a single day' must be interpreted as 'only one time'. Therefore, more than one act of sexual intercourse is not necessary. Furthermore, the use of the word 'stomach' must be read in the same way as our 'the baby is in the mother's tummy'. I collected numerous versions on fecundation and the differences were minor. I believe, therefore, that the stated one fully represents the sense of an event which has been the object of reflection and elaboration in all cultures (McCormack, 1988).

71. Here, the women equate sperm to blood, *sakami*, although in the case of coitus interruptus they use the term *malenga*, water. The two different terms for the same substance, sperm, indicate perhaps two different functions: the first that of nourishing, the second of generating. I very rarely heard sperm being indicated by the term *mele*, milk. The version I collected differs from Rigby's, in which the baby inherits his blood, as well as his physical characteristics, from the father, but which makes no reference to an encounter and combination with the woman's blood (Rigby, 1963).

72. *Munhu na sakami yakwe*: each individual has his own type of blood – this means each individual has his own behaviour and reaction towards *something*.

73. *Misipa*, a term derived from Kiswahili, *mshipa* (pl. *mishipa*), vein. This term is often used by the women to underline the direction, the flow of blood in the veins or the equilibrium in the human body.

74. On this point, some of my informants told me that only one of the three *vidonje* made up the 'baby's house', while the others disappeared, to be discharged during delivery. Another, less elaborate, version foresees the encounter between the male and female seeds as multiple. In fact, it is compared to the three stones of the hearth, arranged at three angles of an equilateral triangle. The male seed encounters only one of the 'female seeds', while the other two close the door, that is, menstruation stops. This is the beginning of a new human being.

75. *Kumosoleza:* to breast-feed the baby while the mother is newly pregnant.

76. The same behaviour, dictated by the same belief that diarrhoea caused by hot milk can be removed by diarrhoea able to eliminate the source of heat, was observed even among the Maasai mothers of Kenya (Patel et al. 1988).

77. In 1933, Cecil Williams (1935) described for the first time the directly observable signs shown by a baby affected by protein-caloric deficiency: scaly skin and depigmentation of the hair and skin. A few years later in 1935, he proposed, in the description of these symptoms, the adoption of the term *kwashiorkor*, from the language of the Ga of West Africa, where he had first encountered the problem. This population used this term on referring to the conditions of a child following the birth of a sibling.

78. All the children born to a married woman, whosoever the parent may be, are children of the legal father, the man who through marriage has obtained a monopoly on the procreative ability of the woman and for which he has paid the bridewealth. Social paternity is therefore never questioned. See Rigby (1969b, Chapter VI); Radcliffe Brown and Forde (1950); Goody (1973).

79. The expression *mbuya ya mlomo,* which cannot be translated literally, sounds more or less like 'mouth lover', that is, the man may visit his lover, but just to chat and for nothing else.

80. As Rigby pointed out (1963) such an institution is present in other areas of Tanzania, although with variations, while it is very similar to that described by Schneider (1961, quoted from Rigby, 1963) among the Nyaturu, a group whose history has often interlaced with that of the Wagogo (Mnyampala 1954). In the Gogo tradition, the term *mbuya* assumes other meanings, for example ancestor, often founder of the clan, whose lineage relationship cannot be proven genealogically, or a very close friend of the same sex.

81. My assistant said: 'Nowadays means to show, to make known, as the case of husband and wife having a lover; the husband, for example, tells the wife that 'Agnes is my friend' so the wife *knows* that her husband has a woman friend.'

82. The entire therapy is motivated by the need to cancel the abnormal physical condition (alteration) created in the woman's body due to her sexual activities and, in the baby's body, due to his having sucked his mother's 'bad milk'. The concept of cooling, (to be cooled, *kupozwa,* from *kupoza*) thanks to a bath of the whole body, *kihovuga,* or of a particular part, *kwizugusa,* can be found in other different circumstances of the lives of the Wagogo (Rigby, 1966a: 7–9). Such behaviour confirms the safeguard, or restoration of a good ritual state, *mbeho swanu.* It must be remembered that *mbeho* may also assume the meaning of 'fresh', of 'the right body temperature'.

83. In the whole procedure, we find the characteristics of the therapeutic process highlighted by Murray Last (1993: 654–5): 'diagnosis and therapy are an analytically distinct process (though they may empirically be one process), ... Diagnosis is concerned with divining the cause of the illness, and therefore works primarily with the past; therapy is concerned with curing the symptoms. Causal explanations, therefore, are not necessarily an accurate guide to clinical practice, not least because the patient need not be the same person as the victim of the agent causing the patient's illness'. This is the case in question, where the child's bad health is caused by the bad behaviour of one of the parents.

84. This onomatopoeic expression, *pwa-pwa,* reiterated several times with the typical tonality of African languages, tries to imitate the sound of continuous diarrhoea.

85. As I have said elsewhere (Mabilia, 2000), in the malnutrition unit of Dodoma Hospital, more than 40 percent of the total number of deaths occur within the first forty hours of hospital stay. The high number of baby deaths seems to be the most concrete response to the idea that the mothers are where they should not be with their babies, in the *wrong* place. The data relative to mortality and to the number of mothers who ran away from the unit in 1999 are as follows: 628 admitted; 98 taken away; and 151 deaths in the first 24 hours after admission. The previous year, the number of children taken away by their mothers was 210 out of 663; 194 deaths of which 71 within the first 24 hours (Serventi, 1991).

CHAPTER 5

MATERNAL MILK:
INDICATOR OF 'GOOD MOTHER'

Breast feeding: a bridge between the different levels of the Gogo social system

The route which has been delineated so far confirms the complexity of breast feeding, its being, therefore, a physiological process steeped in cultural, social and psychological instances which see a woman's actions responding to introjected expectations, formal and informal, day after day, through community life. As a mother, she who is asked to breast feed and rear the children, to be she who, for many months, gives her breast milk, a nutrient which is the artificer of her very own survival, to her own child, her person is invested with a series of responsibilities, made object of a series of attentions, for the interlacing of different types of ties – physiological and social – of which she herself is considered the bearer. The rules which define the nutritional model, that is, what she believes to be right or wrong for her child, the relations which she maintains within and outside the family, have consequently become the ways by which she outlines the contents of a female reality, by which the woman, mother and wife must measure herself during the whole period of breast feeding.

The most significant elements in constructing the identity of an adult male have emerged, comparatively, in his double role of husband and father. In this sense, the disparities which have surfaced between the obligations of a woman and a man and the different responsibilities and attentions of which they are made object should not be surprising, especially in the light of the different tasks which they are asked to perform, as mothers and fathers.

Looking at breast feeding in this sense, the mother–newborn dyad has come out of a truly false isolation and been projected towards the inside of a true system in which actions and reactions participate in community dynamics, defined on the basis of habitual norms. Rights and duties, obligations and prohibitions have become the junction of the weaving of a net of relationships in which men and women move and on which both define expectations and hopes.

Here, therefore, for a mother to be a nurturer, with the weft of ties of different nature – physiological, cultural and social – of which she holds herself and is held to be the bearer, is much more than just the exercise of the potentiality typical of the female body. It is the realisation of behaviours and strategies, where knowing how to conjugate the duties of a mother, of a nurturer, but also those of a wife, becomes of primary importance in avoiding risk to that vital fluid, her milk, the indispensable nutrient for the correct growth of her offspring. This is even truer, we know, if referred to the post-partum taboos that a woman must observe through a long period of sexual abstinence. It is therefore interesting to reflect on the woman as the centre of a series of dynamics, outlined and interrelated by those roles – wife, mother and nurturer – which she must play once married. In attempting to make the ties, overlapping between the different roles emerges, fraught with the possible tensions and conflicts. A woman's behaviour during breast feeding therefore assumes more articulated and precise outlines.

I will begin by considering the woman in her role as mother, closed in the nurturer-newborn dyad, then as wife-mother in the context of *gender*.

The woman: mother and nurturer

The importance of the mother's milk in bringing up offspring is noted. Having indicated it as fundamental until approximately the ninth month of life, *exterogestate fetus,* the period in which the breast assumes the function of the umbilical cord and of an external placenta, emphasises the importance that Western medical science attributes to breast feeding (Bostock, 1962; Woodruff, 1974; Jelliffe and Jelliffe, 1978, 1986; Taylor, 1985; Mohrbacher and Stock, 1997). The Wagogo consider it just as important, although they follow different routes.

First of all, the primary role attributed to a mother is looking after her children. The nurturer–newborn dyad in the Gogo reality is understood as a couple capable of realising a specific nutritional technique, entrusted to the woman's biological make-up and to the child's needs, the results of which fully satisfy the baby's nutritional requirements in the first months of life and make the woman feel suitable for the task she is entrusted with as a nurturer. From this point of view, giving her own milk is therefore understood and experienced by the mother as a *natural* task – a task which confirms her as female-woman, demonstrated by her reproductive capacity, and as woman-mother, demonstrated by her capacity to nurture her newborn baby with only the resources of her organism. These peculiarities of her physiological make-up, however, are affected by convictions and behaviour having symbolic meaning and moral weight, which trace the outlines of a nurturer model in which duties, expectations, emotions, sentiments, joys and worries may interfere with the same quality of her milk.

In this diversified picture, to breast-feed, to give, that is, one's *own* milk to one's *own* child, becomes, for a Gogo mother, to give one's *own* 'good milk', in order to allow one's *own* baby to grow strong and healthy, the condition which is necessary for her to consider herself and to be considered a 'good mother'.

Faced with these different levels, from the physiological to the nutritional, from the psychological to the cultural and social, in which the mother and her newborn baby are so intimately involved, I asked myself whether, in giving, in offering her own breast to the baby, a gesture which is deeply introjected through the inculturative process and experienced with such spontaneity and obviousness that it is considered *natural*, it is possible to trace the dynamics of gift.[1]

To recall the concept of gift as delineated by Marcel Mauss in his famous essay which appeared in 1924 (Mauss, 1965) as a gesture, an action repeated in time, as is breast feeding, may seem a contradiction. How to reconcile the spontaneity, the gratuitousness of a gesture experienced with so much naturalness, with the obligatory nature of gift, as delineated by the French scholar?

To attempt to answer this question means to consider the significance and values that women attribute to maternal milk, so much so as to make it an 'object-good', even better, 'supreme-good', of the bond uniting a mother to her little one. It makes perfect sense however, beyond the same relationship from which the gesture of *giving milk* begins. This is due to the fact that the behaviour and the relations that a woman maintains, not only as a mother and a nurturer, but also as a wife, interact with the qualities attributed to her milk, so much so as to put the favourable progress of breast feeding, and indeed the child's health itself, in danger. Maternal milk, therefore, as Gogo women know very well, is always on a tightrope between being a 'good nutrient' or a 'bad nutrient', in an ambivalence which cannot but call to mind the ambivalence of the term *gift*, at the same time gift and poison, mentioned by Mauss on referring to the Germanic languages.[2]

On answering the question whether breast feeding may be configured as a gift, I will look for those bonds, starting from the *natural* bond which constitutes itself between a mother and her baby at the time of birth – a follow-up of that constituted during conception and gestation – , which make the dyad participate *in the totality of society and its institutions*.[3]

'Good nurturer' and 'good mother'

From the physiologist's point of view, a mother, on giving her own milk to her own child, initiates a circular process in which the frequency and the duration of the stimulation exercised by the baby when sucking the mother's nipple interferes with the psychophysiological reflex of the secretion of milk (let-down reflex). The same breast-feeding modality effectuated by the Gogo women – prolonged, intensive and frequent – promotes the secretion of prolactine and then the production of milk. Breast feeding therefore, develops within a circularity where:

(a) the mother gives her own milk to her child, offering him her breast;

(b) he child receives milk by sucking the maternal breast; this latter action

(c) gives the mother the possibility of giving her own milk again and

(d) allows the child to receive it again.

This reading confirms that breast feeding is a relationship that is fully experienced as a prolongation of gestation, in which the mother, as the experts say, after having donated life, donates her own milk, to continue guaranteeing extrauterine life to her child. At the same time, the close mother-newborn baby interdependence is proposed as an act, which in the circularity of giving and receiving milk in order to have milk again incorporates that mechanism of giving, receiving and giving back which is the foundation of Mauss's study on gift.

This synergy, reccurring at every feed, is not exhausted, as we have seen, as a mere physiological mechanism. It is enriched with cultural, social and psychological meanings and goes well beyond the nutritional value of the gesture, to nurture (allow me this play on words) the sense itself of the relation-exchange between a mother and her child. The two subjects involved are not engaged merely in a relationship which sees them in symbiosis, in a game of mere interest and more or less evident convenience, all within the physicality of their respective physiological needs and potential. They are not engaged in a merely instrumental relationship, in which it is their respective tasks which are more important than the personality of the subjects. There is, in breast feeding, an inherent relationship which is totally cultural, woven with values, affections, emotions, expectations, ways of feeling and participating that deeply tie the two subjects, mother and newborn baby, one to the other. There is a relationship between a mother and her baby that begins from a sentiment of *aimance*, as Caillé defines the field of love, affection and friendship in the sphere of 'primary sociality' (Caillé, 2001).[4]

We know that the physiological and affective dynamics of this reciprocity are not exhausted within the nurturer-newborn couple. The result of breast feeding, in fact, begun on a biological-nutritive plane – giving 'good milk' in order to have 'good growth' -, rich with cultural and emotional instances, completes itself, being enriched with sense on the relational plane in the correlation:

'good milk': 'good nurturer' = 'right growth': 'good mother'

This consequentiality, which may be read as the '*motive* of the gift' to use Godbout's expression (1998:10), indicates not only success for the woman's qualities as a 'good nurturer', but also her capacity to maintain the good qualities of her own milk, a task which belongs to the 'good mother'.

Things do not always follow this pattern, however. During the long months of breast feeding, maternal milk may be subject to modifications and, sometimes, to true alterations which, as we have seen, are attribut-

able to various causes.[5] Casual events and actions conducted with evil
intent by bad and envious people (rarely, however, aimed at harming the
baby, even indirectly) are frequently deemed by women to be due to the
mother's behaviour and, above all, to the situations which she must deal
with as a wife.

Being a 'good mother' – the sum of different instances[6] – is translated
by the woman into an image of herself in the role of nurturer and mother
(highly gratifying on the strictly personal level) and, at the same time,
multiplier of relational richness, for the consensus which she feels growing around her (no less gratifying on the merely individual level as a
woman). This explains the pride with which certain mothers showed me,
almost with ostentation, their small child in good health.

So, to return once again to the concept of gift, if we agree with the statement that it is to be understood only in relation to the meaning that it
assumes in a relationship, giving milk for a Gogo woman has sense when
translated (a) in the good health of her child; (b) in her self-esteem as a nurturer and mother; (c) in satisfying the family's expectations, first of all those
of the mother-in-law and the husband, then of relations in the widest sense
starting from her original family, the neighbourhood and the whole community. It is a return, therefore, in social terms of consensus multiplication.

It is precisely when looking at this consensus multiplication and, therefore, at the strengthening of the network of personal and interpersonal
relations, that it makes sense to interpret breast feeding as a gift. We can
therefore apply here what Caillé said with regard to gift as the centre of a
network of circulation goods and services, the network of sociality:

> In this third network[7] goods are put to the service of the creation and con-
> solidation of social ties, and what is important first of all is not so much the
> use value or the exchange value, but the value that can be called the bond-
> ing value. (Caillé, 1998: 9, my English translation)

Breast feeding, therefore, with its repeated giving in time and its responding to a material need – nurturing the infant with a food that is appropriate for him – with its ideal moments in which the emotional states of the
mother and the child are woven together, makes total sense within the
composite dynamics where maternal milk, the object of giving, becomes
a good at the service of social ties. Therefore, to give one's own milk is a
social action and is viewed as such, 'for its symbolic nature, held to mean
actively and indissolubly mixing obligation and freedom, interest and dis-
interest' (Mauss, 1965: 228; my English translation).

Mary Douglas's statement in the introduction of the French scholar's
essay, 'the free gift does not exist' (Douglas, 1989), synthesises well the
sense of a nonproposable gratuity referred to the dynamics of the relationship created by the act of *donating*. This relationship, in the case of the
object, is not concluded only and necessarily with giving, receiving and
giving everything back within the nurturer-infant dyad. The gesture, on
assuming a circular form, extends itself in the same way as a circle which,

formed by a stone being thrown into a pond, multiplies until it covers the entire surface of the water. For a mother, to 'give milk' to her own child means tying *other* subjects to them, involving *other* bonds, giving value to her capacity to confirm, reinforce and make relationships.[8]

The gift of 'giving milk' has, therefore, an intrinsic propulsive capacity which overrides the dualism of the gesture, to involve the social environment inside which the gesture itself takes form. It is by raising one's eyes and looking at others, looking at the circles forming after having thrown the stone, that giving one's milk, for the Gogo women, assumes the character of a gesture serving both life and ties. This adds value to the nongratuity of the dynamics of the gift, in the measure in which it underlines the importance for a woman in her role as mother-nurturer.

Hélène Cixous, on tracing the border lines between a female economy and a male economy, believes that the gift is an instrument used by women to establish relationships. In consideration of this, she recalls the most peculiar of maternal gifts, a mother's love towards her child, the gift of her own milk – comprehensible both outside the patterns of economically quantifiable exchanges and the laws of economic compensation which hold up mercantile exchanges. She also considers the gift nongratuitous, not free, but there is no contradiction – what distinguishes it is its *why* and its *how*. The difference lies in the values which are affirmed with the gesture of giving, the *motivations* which are called into question, the *type* of restitution that the giver hopes to obtain and the *use* that derives on the interpersonal relations plane (Cixous, 1997: 159, 163). All of this can be traced in the *natural* gesture of giving one's own milk in Gogo mothers – a gesture which summarises composite and comprehensible values well beyond the dual nurturer-infant relationship, to make reference to that network of sociality in which, for a woman, the roles of mother and wife find expression and merge.

What happens when this network of sociality goes into a crisis due to the failure of breast feeding, so much so as to jeopardise the healthy growth of the child? What happens when the maternal milk becomes a danger[9] to the health of the infant, so much so as to make the 'good mother' a 'bad mother'?

In order to answer these questions, we must turn to the woman, to the weave of bonds and relationships, to the rights and duties, to the rules of behaviour which can, and in fact do, enter into conflict when the mother-nurturer priorities come up against those of the wife-mother.

From 'good mother' to 'bad mother'

We have seen that there are circumstances of different nature and gravity which may interrupt the success of the circularity represented by the giving of 'good milk', aimed at obtaining a 'correct growth', synonym of the 'good nurturer' and of the 'good mother'. The changes – or rather the possible deterioration – that we are interested in here, affect the quality of maternal milk (the consequence of a mother not having observed postpartum sexual abstinence) due to the close tie existing between her body, her milk and her child. This means that the dependence between her and her child forces the woman into the management of her person where, for long periods of her life, her priority needs to be to fully satisfy the development and well-being of her last-born child. In other words, by imposing sexual abstinence, a woman's duties as nurturer are put before those of wife and mother. This priority is not painless: an individual cannot be separated from her roles.

The intellectual capacities and the equally distinctive experiences of a subject, whether male or female, in perceiving and introjecting the cultural patrimony of the community within which he or she grows, work towards providing a 'sense of being', a 'sense of being in the world' which is at the same time shared in its *guidelines*, as defined by Helmann (1990: 3), but experienced in a personal way, in the most intimate feelings of each individual.

Anthropological research has often stressed how in many traditional societies, it is not the single individual who occupies the centre of a series of relations, but it is the relations between individuals which are placed in the centre and which define each individual, according to the roles that they assume within the community of which they are a part. This does not, however, stop an individual from presenting himself with a particular character, with his own specific *individuality*, recognised by the community, and which makes him live, in 'his own very personal way', what has been transmitted to him.

This means that every Gogo woman, just like every other human being, is a person with a well defined individuality, with her own way of thinking, of feeling, of perceiving her legitimate behaviour, with her own expectations and her own sense of duty. She has *her own style* of behaviour, her own style in presenting herself, of proposing herself, her own style in relating to others. It is, in other words, *her own way of being*, between an *'ideal how'* and a *'real how'*, that makes her a unique and unrepeatable person.

If, as Marc Augé said: 'individual beings do not exist if not through the relations which unite them. … the individual is nothing but the crossroads, necessary but variable, of a totality of relations' (Augé, 2000: 24; my English translation), then in this relating to each other, the individual may find himself assuming many identities,[10] which can enter into conflict one with the other, promoting tension in the subject himself and in his relating to others.

This is because, in the interlacing and intersecting of relations, culture, together with ideals and symbolic meanings and values, acquires visibility within different societies, thanks to the forms through which individuals organise their own associate life, through the concrete actions of their members. It is in *doing* that we meet the inevitable overlapping of individuality and identity. Here, I mean to refer to the co-presence of levels and models of participation in community life which live together, crossing each other, in every single individual.

In these dynamics, furthermore, being born male or female acquires its own value, becoming the starting point from which gender belonging is defined and makes sense. *Gender* is a strong analytical and propositional concept which allows us to see a woman, and a man, as a product of a process in which ideas, norms, behavioural models, relationships, expectations and visions of the world go to form females and males, differentiating them, first of all on the merely cultural plane, and then on the social and political planes. It is the process through which, as Picone Stella and Saraceno recall on citing Gayle Rubin (1975), 'every society transforms biological sexuality in products of human activity';[11] it is that concept referring to the totality of processes put before the 'social construction of the male and the female' (Picone Stella and Saraceno, 1996: 7; my English translation).

Everything is interrelated in this construction, at times in antithesis, at times complementary, at times hierarchical, in a mixture of elements towards which every individuality, extinguishing itself, seeks integration and harmony. It is done by implementing strategies and choices, sometimes on the borderline, sometimes overstepping the demarcation line between 'right' and 'wrong', between 'permitted' and 'forbidden', in order to set out on new roads, the prelude to future changes. This processuality 'presents a bill to be paid', on both the psychological and relational plane.

It is not surprising therefore, if, in having to manage her own person, even an adult woman in the Gogo community must try to move with ability and determination between expectations and hopes, hers and of others, between the rights and the duties she has been taught, in the implementation of her roles as wife, mother and nurturer. These, because of their co-presence, are not always *consonant* one with the other.

If what the community expects from a woman is for her to guarantee a 'living together' of her duties as a mother-nurturer, dedicated entirely to her infant, and her duties as a mother-wife, subdivided between looking after the *nyumba* and the rights of her husband, then the task is not easy for her. As a mother, she continues to look after her other children, carrying out the tasks entrusted to her, starting from the search for and the preparation of food, while as a wife she must, principally, stand up to her husband, refusing his ever-increasing pressure on her to have sexual intercourse. When she ends up by giving in to this pressure, thereby going against the rules, she knows intimately that her motivations go beyond a pure and simple giving in to her husband's insistence. Giving in can also mean containing her jealousy as a wife who is trying to stop her husband from going with other women. However, she may also give in to her desires as a woman, legitimately perceived, thanks to an educational process which traditionally intended to regulate, and not deny, the force of the libido.

Her duties as a nurturer and as a wife and her most intimate and private impulses then clash in an unequal fight, faced with her husband's demands for his rights, or with her possible jealousy, or with her woman's sensuality, or, perhaps, a little of all of these. This is the deep meaning of

the words of the older women when they said that a mother 'is not able
to deny herself to her spouse', words which, at the same time, indicate
failure to observe another rule that, requires that the relations between
spouses be free of passion and jealousy. According to tradition, relations
between husband and wife must not be governed by sentiments, by
strong and disruptive impulses that may cause possible conflicts. These are
the ways of living a relationship typical of lovers, where even magic may
be used, if necessary, by both parts, in order not to lose, or to conquer the
other companion. It is important, then, to avoid this accumulation of
emotions to safeguard, with controlled, calm and formal conjugal relations, a tranquil *ménage* and, with it, the ties of affinity still considered precious in Gogo customs.[12] Certainly, in choosing a spouse, physical
attraction and amorous sentiments are not, and have never been,
excluded, but a young woman's good character, her ability and constancy
in her work and her potential reproductive capacity continue to play a
major role.

If passion was to be relegated to the particular relationship with the
mbuya, as a viaticum for a more stable conjugal relationship, today, with
the decreased incidence of this figure, together with a generalised loss in
the efficacy of the habitual norms with regard to sexuality, what was once
the typical behaviour of lovers now seems to characterise, more and more,
the relationships between husbands and wives.

When the result of this warring, between prohibitions and desires and
between consensus and submission, is a new pregnancy, a mother feels a
deep unease and growing anxiety. Firstly, because of the signals that she
begins to perceive in her own body and which she tries to deny to herself,
then because of the pervasive sense of guilt she feels when her baby's
diarrhoea leaves her no further doubt or alternatives. She cannot, in fact,
wait any longer: she knows that the changes in her body, due to a new
pregnancy, have transformed her milk. The infant already born can no
longer continue to suck and, therefore, his correct growth is in danger.
Hence, everything that she had obtained and that was recognised as hers
up to that moment, as a 'good nurturer' and a 'good mother', is compromised because of her behaviour. The 'good mother' becomes a 'bad
mother', she who in not observing the obligations imposed on her by her
state of nurturer has made the transformation of her 'good milk' into 'bad
milk' possible, with all the consequences which she well knows, on the
physiological level for her child and on the social level for herself.

The community's reproof is all on her, sanctioning a gender disparity
which sees her by far the most disadvantaged compared to the husband-
father. On the other hand, she is the one who must take care of the offspring and therefore, she must know how to manage her own body as the
rules dictate.

If her sense of guilt and the community's consequent reproof are
sources of shame, at the same time, they may be two strong stimuli to
seek for a solution to such a shameful situation. We have seen, in fact,
that, in the case of a precocious pregnancy the image of the 'good mother'

in the family and the community may be recovered by: a rapid distancing of the breast from the child; an equally rapid treatment to eliminate the bad milk from his stomach, the cause of the diarrhoea; a better and continuous maternal commitment to the child's diet, as the mother can no longer breast-feed. It will not be an easy task to offer her child an adequate diet, both for environmental reasons and because of cultural concepts and behaviour,[13] but by trying hard, a mother may find self-esteem once again, and regain for herself a positive image in the community, with all that this means on the level of *relational wealth*.

Among the new generations, according to the older women, sexual abstinence during breast feeding is not rigorously followed between spouses, as tradition demands. However, these are changes which, at the time of my research in the field, lived side by side with tradition, in that play between past and present where the *dissonant* behaviours[14] of women or of men, or of both, seem like challenges. By deciding their own present some women contribute, many times without knowing it, to promoting the future direction of the community.

Some statements collected from the younger mothers suggested that some discordant, some *dissonant* individualities, were making headway, marking with their ideas the outlines of new and original styles, a warning of a process of change, or at least of a weakening of the rules of this private and intimate theme.

Statements like: 'Why should I be ashamed of being pregnant? I have a husband and I can't deny myself to him. Then, the father cannot damage the child!', or 'The important thing when you sleep with your husband, is to be careful not to become pregnant', clearly show that a new way of understanding sexual relations between the parents of a newborn baby is emerging. It does not, however, push as far as the rule forbidding breast feeding during a new gestation. Only one young mother told me that she had breast-fed during the first months of pregnancy without harming her last-born baby. Perhaps it was the baby's good health which made her continue breast feeding!

Talking to a few young mothers, I was able to ascertain that it is not, in fact, so much the bond between the parents and the newborn baby, created at the moment of conception[15] that is in question, but that this bond may be altered, deteriorating the quality of the mother's milk, due to sexual intercourse with the husband-father, unless pregnancy is the result of the same.

If being pregnant during breast feeding is still a transgression which is the object of reproof and blame from the community, to have sexual relations with men other than the husband is a violation which becomes a source of social tension, and goes far beyond the complex elements present in a precocious pregnancy.

'Bad mother' and 'bad wife'

An adulterous woman not only clearly proves that she has not considered the behavioural rules which aim at disciplining her sexuality as a nurturer, clearly showing that as a mother, she does not take care of her children's health, but also show that, as a wife, she has not fulfilled her duties and rights towards her husband, having allowed another man access to her body.

When Rigby faced the problem of extraconjugal relations in the field, running into the figure of the *mbuya*, his informers traced a picture in which, as they stated, it is above all among the young that sexual rules, *moto wuwo* (lit. this fire), were already declining in favour of unstable relationships which were not regulated. Morality was changing and relations were assuming more and more the connotations of 'theft', *kuhiza*,[16] not a new situation but one which was causing blame and worry because of its accentuation.

As in Rigby's time, the older women judged the younger ones severely, attributing to them senseless behaviour, or, at least, nonrespect of traditional morals:

> They don't care about the rules, they don't know how 'to say no', giving themselves for a piece of soap, *sabuni*, or for a drink of beer, *ujimbi* (traditional beer), or for a dress, *mwenda*. They are prostitutes, *msenhya*.

And the result of all this laxity? Today there is a great mixture, *mhanga-hango*, and so you see children who have difficulty in walking, who have weak legs, *vigulu vininyinago!*', is what the women told me during a long conversation about the health of the children in the village, confirming just how much they considered a mother's behaviour, as a nurturer, linked to the health of an infant.

Naturally, this way of understanding and interpreting the present by the older women reflects the unease of individuals, whose youth was experienced in a different context and who tend, therefore, to judge severely those changes which no longer correspond to their behavioural models. In general, as people get older, they find it difficult to adapt to new ways, considering it a loss, a failure, to respond more fully to the community's needs.[17] However, 'my' older women were, hypothetically, Rigby's younger women, a generation which, to hear the older people of that time, were going against traditional moral principles!

The situation with regard to sexual customs is particularly complex and difficult to define, here as everywhere. To label an aspect so intimately connected to the psychology of each individual, as merely *changed* or *in the process of changing* may not correspond to people's experience, where past and present penetrate each other, live side by side, oppose each other, change each other, in a continuous play of alternation. It did not surprise me, therefore, when a young mother, in confirmation of a situation in which tradition and change are measuring their respective convincing strengths, stated:

> Once upon a time, a mother did not have a lover during breast feeding, but today it is not like that. It seems that some cannot do without. When I got married my mother said to me: now that you are a wife, you have to forget

about any man outside marriage, completely, *lukulu*, while you are breast feeding. Stay away from men who want to give you presents to show you 'something'! I followed her advice and my baby is growing up without problems. Some of my acquaintances have not followed the rules taught us by the older women, *wadala*, and so their children are not growing well.[18]

These statements only confirm a present which is on a tightrope between tradition and change.

The figure of the *mbuya*, in its most traditional form, as has already been said, has lost much, if not almost all,[19] of its important function on both the individual level (as a regulator of men's and women's sexual life), and on the level of kinship group – so much so as to be considered complementary to marriage.[20] However, the women confirm the importance of norms, which although today seem to be transgressed with a certain frequency, are not violated without reproof. Furthermore, on the strictly personal level, they testify, for a man, the value of having a lover to prove one's virility and sexual ability and, for a woman, the importance of having her powers of attraction, her sex appeal, confirmed.

Yesterday, as today, for a man to be a lover, *mzelelo*, and for a woman to be a lover, *muhinza*, signifies a relationship woven with emotions, passion, jealousy and sensuality culminating in the sexual act, where the search for pleasure is also the result of being able to give pleasure. My interlocutor's words, however, suggested also a new way of understanding extramarital relations where not only sensuality and passion dictate, but also the need to have or the pleasure of having 'a new dress or a *kanga*' or 'a few pieces of soap', in exchange for sexual intercourse. It is these 'new desires'[21] which motivate women's willingness to have occasional lovers and the consequent dangerous mixture, *mhangahano*.

Faced with this licentious picture, in the general condemnation of adultery it is the woman who is always the main scapegoat. She becomes more and more the object of reproof if her going with men other than her husband occurs in that period in which, as a mother, her duty as a nurturer should disciplin any other desires.

The suspicion of a mother's adultery begins to spread among the women living around her, starting with the mother-in-law, as they observe the changes in the baby's health. It is a diarrhoea that resists all treatment, traditional or otherwise, and the consequent regression in development, accompanied by a worsening of the child's health, is, according to the older women, the final proof that the mother has had intercourse with men other than her husband.[22]

The woman who is the object of these accusations may experience the situation on three possible different levels:

(a)   she knows that she is not pregnant and that she has not violated any taboos, but her husband may have done so, or he may be the object of a spell;

(b)   she knows that she is not pregnant, but that she has violated a taboo, giving herself to another man;

(c)   she knows that she is pregnant and that she has violated a taboo;
       it doesn't matter with whom: the father of the child in her womb
       is still her husband.

The last of the three situations is, as we have seen, dealt with according to
tradition: the mother responds to the extraordinary cause provoking the
infant's diarrhoea with a specific and rapid therapy, committing herself to
looking after and feeding her infant with those weaning foods which are
available, as he can no longer take her milk.

Case (a) is the most dramatic, because she knows, on the one hand, that
witchcraft is rarely brought into question when directed at a child and, on
the other, that it would be difficult to involve her husband as the person
responsible – firstly, because his period of abstinence is more limited,[23] secondly because it is extremely arduous to investigate his sexuality.

When the woman knows in her heart that she is responsible for her
child's condition – case (b) – having violated the post-partum abstinence,
everything becomes more serious for her, starting with the search for a
remedy aimed at curing his condition. As the women say when reasoning
on this latter case, she wriggles out of the search for the 'right cure'. The
woman, however, does not remain helpless: when the diarrhoea starts she
follows the usual procedures, starting with domestic remedies, and, if
there are no results, she turns to an expert. As the days go by, however,
and on seeing that all the remedies attempted give no results, her anguish
increases. While the failure of the attempted treatment worsens her
child's health, it is also more difficult for her to dissolve the doubts about
her behaviour, believed to be the cause of her child's condition, and the
pressure of which she feels more and more upon her.[24]

The question is: although the mother is concentrating on looking after
her baby, why does she not look for the 'right remedy', as the other
women would like? Why is the search for the *mapande*, that special therapy which will cure her child, continually put off and, in fact, avoided?

The answer is to be found in the intimate conviction, as implicit as it is
heavily felt by the mother and which, at the same time, she continues to
deny: her admission of guilt for the deteriorating conditions of her child's
health. It is this confession, firstly to herself, and then to the healer, which
delays the use of the 'right' remedy, taking a mother, the women say, to
the 'wrong' places, with grave outcomes for the child's life.

It is not difficult to understand why the admission of guilt is translated
into an admission of triple failure: as a 'good nurturer', 'good mother' and
'good wife'. She has failed as a 'good nurturer' because her milk, of good
nutritional value and dedicated to giving life has become, due to the fact
that she has permitted a mixture to occur in her body at the wrong time,
a bad nutritional element, able to kill her child – she is a 'bad nurturer'.
As a 'good mother', she feels the whole weight of her defeat with regard
to her child and the family, sensing a loss of credibility and consensus
which makes her feel entirely alone. The relational richness, which she
has constructed and maintained through her total dedication to her child,

clearly proved by his good development, has now collapsed – she is a 'bad mother'. As a 'good wife' she knows she is now the object of accusations, sometimes accompanied by violent reactions from her husband, and which, if due to their gravity also involve her original family, will only increase the tension and the recriminations. She is thus a 'bad wife'.

The circumstances which a woman is forced to deal with witness not only her guilt for not having observed the rules and the weight of the effects of these cultural dictates which mark her behaviour as a nurturer, a mother and a wife, but also very strong psychological inhibitions which condition her actions as an individual. How can it be a surprise, therefore, in the light of these considerations, that the woman does not go in the 'right direction' and with haste, which is what the other women reproach her for? How can it be a surprise if she tries to remedy the progressive worsening of her child's health with traditional solutions or by going to experts, in order to prove, at the same time, by the improvement of the child's health, that she has not violated the rules?[25]

Disregarded duties and obligations, sentiments, emotions and growing fears, all contribute, therefore, to defining her behaviour at a time when her relations with her kindred group are becoming more and more strained because of her child's health. This is because, should she find herself pregnant during the period of breast feeding, she would be altering the 'natural order', as is well summarised by one woman's words on the unfeasibility of a condition which sees 'a mother carrying two weights, one on her stomach and one on her back';[26] and if it is the rule of adultery which has been violated, she would be altering the 'social order', for having decided autonomously on the use of her own body. Pregnancy, in the eyes of the community, represents legitimate sexuality, while adultery means illicit sexuality.

In the light of anthropological literature, we know that despite the wide variety of social and cultural meanings attributed to the sexual and reproductive sphere in different sociocultural contexts, it does not escape careful regulation and control.[27] The result is that the exercise of one's sexuality is never a private and intimate matter, involving the two partners alone. This is even more evident in those societies, like the Wagogo, in which kinship ties play an important role in the functioning of a community.

The Wagogo have high consideration for the ties of affinity which are constituted through the marriage contract, so much so, as to safeguard them by educating towards a marriage bond not based on passion. This marriage contract sanctions a number of rights and duties for the future spouses, between their respective families through the definition and payment of the bridewealth. With these understandings, adultery represents a woman's failure to observe her husband's rights and also, therefore, nonconsideration of the risks that the affinity ties may run due to her inconsiderate behaviour as a wife. It does not signify here facing up to a pregnancy in the wrong moment, nor is the husband's fatherhood placed in doubt. Payment of the bridewealth means that a man has acquired the right to the fruits of the reproductive capacity of the woman who is his wife: *pater* and *genitor* of the children that the spouse gives birth to is always the husband. What is being questioned here is the fact that the wife has given access to 'another' man, exercising a right over which she has no title. The fact of having conceded herself to 'another' man means that she has violated the husband's right to dispose of access to her sexuality, of which he is the sole holder. This right must not be confused with the right over the use of her body, as this may be accorded temporarily, as we have seen in the tradition, by the husband to another man, *mbuya*, or agreed upon in partner exchange, *kusutana*.

Adultery, however, does not only represent violation of a husband's rights and therefore a woman's failure to comply with her duties; it also shows, when it occurs during the period of breast feeding, her failure to comply with her duties as a mother and nurturer towards her child, as she has altered the 'object-good' which in the synergism of the described exchange – giving milk, receiving milk, to have milk again – offers the infant the greatest part of his chance of survival.

These two cases, pregnancy and adultery, two grave violations of post-partum taboos, with their equally serious repercussions on the infant's health, are two key points in not only understanding a mother's behaviour during breast feeding, but also in explaining gender inequalities. In fact, if every time a serious change in the quality of maternal milk, demonstrated by the infant's suffering a serious form of diarrhoea, is attributed to the mother's incorrect behaviour, then maternal milk becomes the 'object-good',[28] thanks to which the behaviour of a woman, wife, mother and nurturer can be controlled. This occurs at a time when the control of sexual behaviour has become fragile and less supported by proven automatisms.[29]

If the traditional dynamics which controlled sexual behaviour have lost most of their efficacy, then other guards must be found. What better controller of the use that a woman makes of her body than her own milk, given that, as a mother and a nurturer, it must serve the survival needs of a creature as fragile as it is precious? What better deterrent against the dangers a mother's milk may be submitted to, following behaviour which seriously compromises her role as nurturer, mother and wife, in controlling a woman's sexual behaviour?

A woman's sexuality, her reproductive capacity and the form and contents of her relations become the centre of social control – a social control which defines various periods of post-partum abstinence for men and women, different consequences for the violation of these rules, different psychological responses from men and women and different expectations from the community.

A brief return to the village

On finishing my research in the field, I found that many questions had accumulated to which I would have liked to find an answer. In the same way, at the end of my route, my analysis of the way in which breast feeding is practised by Gogo mothers opened up new scenarios for investigation and study, in which the phenomenon of changes linked to male mobility no longer connected to pastoral activities alone, the role assumed by women in their absence and the matrimonial and kinship dynamics in general would merit deeper investigation. But, more generally, it would be worth reflecting on the concepts of 'good mother' and 'bad mother' as attributes which invest the woman *tout court* with such a burden of responsibility and expectations which go beyond specific ethnographic traits, rendering the 'Other' not *so* much 'Other'.

Ten years later, I returned to Cigongwe.[30] All the emotions and thoughts which had stayed with me during those years were revived in me during the journey approaching the village. Looking around, everything seemed unchanged, with the exception of a police post on the track just before staring the descent to the village. This novelty, however, had not deterred the community of baboons from their rapid coming and going between the rocks and bushes, nor had it reduced their curiosity.

On arriving near the dispensary, I left the car as usual and made my way, without hesitation, along the paths I knew so well. Everything seemed to be as I had left it, except for a few new huts signifying a probable scission of the *kaya*, or new nuclei coming from other areas, as could be seen from the signs on the ground of huts having been transferred to another area of the village. Moving around in spaces well known to me, I met, one after the other, the women who had contributed most to my work in the field; I saw young men, women and children whom I knew – and noticed some painful absences.

Behind this scenario, however, in which people seemed to move with the rhythm I knew so well, I knew that things were different, *how* and *how much* I did not know. What I sensed, but could not see, was marking the trails from which it would be possible to find a new and different sense in that staying together which the passage of time had started and which the present continues to elaborate.

Notes

1. I have already looked at gift as a possible explicative concept of breast-feeding dynamics (Mabilia, 2001a). Here I mean to conclude those reflections on breast feeding as a complex and composite phenomenon which moved me towards this work.

2. 'The theme of the funereal gift' – Mauss writes, – 'of the present or the good which *changes into* poison is fundamental in Germanic folklore' (Mauss, 1965: 267, the italics are mine). *Gift*: a double meaning, not fruit of chance, as Caillé underlines, recalling the double meaning of the Greek term *dosis*, gift and dose of poison (Caillé, 2001: 262).

3. From the introduction onwards, I have considered the act of a mother giving her own milk to her child as belonging to those social phenomenon '*which put into motion the totality of society and its institutions*' (Mauss, 1965: 286).

4. 'Primary sociality' is that of the family, relations, alliances and neighbourhood and is characterised by *aimance*. Caillé distinguishes it from 'secondary sociality', where 'what counts is the efficacy of our actions, they are the functions we absolve'. Gift, however, is not foreign even to this (Caillé, 2001: 256; my English translation).

5. The inadequacy of maternal milk with regard to the changing needs of the baby, which when growing needs more nutrients, belong to the order of natural development of his organism and do not, therefore, question the nurturing quality of the mother.

6. Confirmation of the initial setting in which I meant breast feeding as a composite and complex process.

7. By 'third network' Caillé means the network of sociality, distinguishing it from that of the market and the public economy. The 'third network' is rarely noticed for itself but it is no less essential (Caillé, 1998). For the third modality with regard to the circulation of goods, see also his 'Notes on the Paradigm of the Gift' in Maraniello et al., (2001).

8. Godbout, by affirming that one-way relationships are not relationships, underlines the importance of the non-one-wayness of the gift. He affirms that a gift is, not by chance, the act of talking, exchanging words, 'the first perhaps' (Godbout, 1993: 13–14).

9. I have spoken various times of the 'risks' of breast feeding, with reference to the possible alterations that the women think maternal milk may be subject to, following external events or the behaviour of the mother herself (Mabilia, 1995; 1999).

10. I use the term 'identities' here in the way it was used by Amartya Sen in Gender and Co-operative Conflicts, (1990).

11. Cited in Picone Stella and Saraceno (1996: 7).

12. Cognatic kinship and affinity ties are still present within the different domestic units, both in terms of cooperation and of choice of residence (Rigby, 1962b; 1966c; 1969b).

13. Beyond what has already been said, for further considerations on this aspect see Mabilia (1996b; 2000).

14. With the term *dissonant* I wish to indicate the behaviours shown by some individuals which do not harmonise with traditional models and which act on the community, promoting reactions, or of praise or of reproof and accompanied, in both cases, by bewilderment for the new scenarios which they reveal.

15. I have already said how little the new generations know about physiological reproductive processes; it is knowledge, however, as I stated in the previous pages, which must be considered specialist knowledge.

16. *Kuhiza,* (to steal) indicates the relationship of two lovers without the woman's husband's knowledge, therefore, the wife's adultery. In this relationship, the 'theft' of the husband's rights over his wife's sexuality is underlined. They are not his exclusively, however, because the cases of *kwilajila* and *kusutana* show that a husband has full possession of his wife's reproductive capacity and control of how her sexuality will be disposed of.

17. Sexual behaviour has been subject to and is experiencing deep changes and it is not easy to evaluate the importance of this in the private life of the single individual. Changes have affected the economic and social fabric of Tanzanian society, even in the rural areas. The problem of the spread of HIV/AIDS is dramatic proof of this, although, at the time of my work in the field, there were only echoes of this terrible epidemic, of something that was coming,

that was being brought from the city. Literature is very rich on the theme of HIV/AIDS. I cite here a few works referring to Tanzania: Quinn et al. (1986); Killewo et al. (1994a,b,c,d); Lwihula et al. (1994); Klepp et al. (1995); Talle et al. (1995); Blystad (1995); Haram (1995, 1999, 2001); Kwesigabo et al. (1998); Kesby (2000); Baylies and Bujra (2000); Kwesigabo (2001).

18. In telling me about how times have changed, my interlocutors, whatever their age, always referred to others as those who were not respecting traditional rules.

19. The consensual and reciprocally noted exchange of partners between married couples is an exception. It was explained to me as an ongoing form of exchange, between two couples, of the right to sexual access, *kusutana* (Rigby, 1963).

20. With regard to this, Rigby's words are significant: 'The close network of kin and affinal ties … is functionally interrelated with the *mbuya* (permitted adultery) relationship: These semi-institutionalised relationships tend to regulate extra-marital sex relations within the local group and the network of kinship, and probably also contribute to the stability of marriage' (Rigby, 1969b: 245).

21. Relationships between lovers centred on the exchange of sexual favours for money or goods is a widespread practise today in many parts of sub-Saharan Africa, but it cannot be motivated only on the basis of desires connected to the possession and use of goods. The motivations are more complex and are related to women's expectations in their role as wives and to their husbands' role, understood on a level of equality and respect, expectations which are not always met. Precarious economic conditions are not foreign to requests which seem to commercialise a relationship in which, however, there may be affection between the subjects (Bledsoe, 1980; Haram, 1995, 2001, 2003; Silberschmidt, 1999).

22. A diarrhoea lasting more than a few days, sometimes accompanied by vomiting, causes loss of appetite and a state of apathy, a lack of vitality in the child which sets off a vicious circle: the lack of appetite reduces the number of feeds, leading to a decrease in available milk. I refer to the description given to me by the women on talking about a child suffering from a serious form of diarrhoea: 'comparing it with a wizened old being or they point out its serious condition saying that "the child has sunken eyes, a sunken fontanelle, its chest runs and has a tail" because its skin shrinks, baring its coccyx, they are talking about dehydration symptoms according to biomedicine' (Mabilia, 2000: 196). This description of the conditions of certain infants refers not only to a state of severe dehydration, but also to health status which is compromised by diarrhoea, the cause and effect of other possible pathologies.

23. If the three or four months of abstinence required from her husband have passed, *kuwika mwiko,* and the rite *kupagata mwana* has been carried out, then the husband is automatically excluded from the issue.

24. It must be remembered that in circumstances like those being described, the woman is always held responsible. It is up to her to demonstrate her noninvolvement with the child's condition.

25. On this point, I have written: 'Cultural, social and psychological factors can induce her to put her responsibilities in second place, ascribing her child's condition to a cruel fate. Through her continuous visits to the *waganga*, she finds a way of justifying and defending herself. … is this not a case of selective neglect? (Scheper-Hughes, 1984, 1987b, 1987c, 1988, 1992). I agree with Howard (1994) that this definition can be misleading. She refers to 'selective

survival', stressing the complexity of the social and cultural context of child care responsibilities' (Mabilia, 2000: 201).

26. See Chapter 4.

27. I have written, with reference to the spread of HIV/AIDS in sub-Saharan Africa, (Mabilia, 2001b), about how the value of organisational and moral principles aimed at the regulation of sexual behaviour in traditional African societies was misunderstood by the West, due to ignorance, lack of interest and prejudice (Beidelman, 1982; Ahlberg, 1991). It is still possible to read articles like those of Caldwell et al (1987; 1989) in which it is clear just how much, – in the past just as today – sexual behaviours were and are interpreted through a 'curtain of prejudice' (Ahlberg, 1991; Le Blanc et al. 1991).

28. I was able in my previous intervention to define maternal milk as a 'barometer' controlling the mother's behaviour (Mabilia, 1999).

29. Together with sexual behaviours, the worsening state of economic and environmental conditions was also evident. Desertification, repeated failure of the rains (in my three years' stay, there was only one regular rainy season) with the consequent difficulties for subsistence agriculture, which together with the different distribution of animals among the *kaya*, as has been seen in this work, were marking a deterioration of the conditions of life. This deterioration was reflected in a worrying way in children between zero and five years of age, an age group which is well noted in statistics on the conditions of infancy in the world, and in particular, in sub-Saharan Africa (UNICEF, 2003).

30. I spent five weeks in Tanzania between the months of September and October 2002.

BIBLIOGRAPHY

The works of Peter Rigby that refer to the Wagogo

(1962a), 'Aspects of Residence and Co-operation in Gogo Village', East Africa Institute of Social Research Conference Proceedings, Kampala, E.A.I.S.R., 1–13.

(1962b), 'Witchcraft, Kinship and Authority in Ugogo', East Africa Institute of Social Research Conference Proceedings, Kampala, E.A.I.S.R., 1–14.

(1963), 'The *Mbuya* Relationship and Marriage in Ugogo', East Africa Institute of Social Research Conference Proceedings, Kampala, E.A.I.S.R., 1–14.

(1966a), 'Dual Symbolic Classification among the Gogo of Central Tanzania', *Africa*, 1, 1–17.

(1966b), 'Sociological Factors in the Contact of the Gogo of Central Tanzania with Islam', in I. M. Lewis (ed.), *Islam in Tropical Africa*, International African Institute, 266–88.

(1966c), 'Gogo Kinship and Concepts of Social Structure', Makarere University Press, 1–14.

(1967a), 'Time and Structure in Gogo Kinship', *Cahiers d'Études Africaines*, 7: 28, 637–58.

(1967b), 'The Structural Context of Girls' Puberty Rites', *Man*, (N.S.), 2: 3, 434–44.

(1967c), 'Local Government Changes and National Elections', in L. Cliffe (ed.), *One Party Democracy. The 1965 Tanzania General Election*, Nairobi: East African Publishing House, 76–104.

(1967d), 'The Tale of the Ungrateful Grandfather', in P. Rigby, and G. Seryagwa, *Transition. A Journal of the Arts, Culture and Society*, 28, 135–40.

(1968a), 'Joking Relationships, Kin Categories and Clanship among the Gogo', *Africa*, 2, 133–55.

(1968b), 'Some Gogo Rituals of "Purification": an Essay on Social and Moral Categories', in E. R. Leach (ed.), *Dialectic in Practical Religion*, Cambridge Paper in Social Anthropology, Cambridge University Press.

(1969a), 'Pastoralism and Prejudice: Ideology and Rural Development in East Africa', in P. Rigby (ed.), *Society and Social Change in Eastern Africa*, Makarere Institute of Social Research, Nkanga Editions, 42–52.

(1969b), *Cattle and Kinship among the Wagogo*, Ithaca: Cornell University Press.
(1971a), 'Politics and Modern Leadership Roles in Ugogo', in V. Turner (ed.), *Colonialism in Africa, 1870–1960*, Vol. III, *Profiles of Change: African Society and Colonial Rule*, Cambridge: Cambridge University Press, 393–438.
(1971b), 'The Relevance of the Traditional in Social Science. On Interdisciplinary Approach to Planned Change', University Social Sciences Council Conference, Makerere.
(1971c), 'The Symbolic Role of Cattle in Gogo Ritual', in T. O. Beidelman (ed.), *The Translation of Culture*, London: Butler and Tanner, 257–91.
(1977), 'Local Participation in National Politics: Ugogo, Tanzania', *Africa*, 47: 1, 89–107.
(1985), *Persistent Pastoralists. Nomadic Societies in Transition*, ZED Books Ltd.
(1987–88), 'Pastoralism, Egalitarism and the State. The Eastern African Case', *Critique of Anthropology*, 7: 3.
(1990), 'Pastoralist Production and Socialist Transformation in Tanzania', in C. Salzaman and J. G. Galaty (eds), *Nomads in a Changing World*, Naples.

General Bibliography

Aborampah, O. M. (1985), 'Determinants of Breast Feeding and Post Partum Sexual Abstinence: Analysis of Sample of Yoruba Women, Western Nigeria', *Journal of Biosocial Science*, 17, 461–69.
Agbasiere, J. T. (2000), *Women in Igbo Life and Thought*, London: Routledge.
Ahlberg, B. M. (1979), 'Beliefs and Practices Concerning Treatment of Measles and Acute Diarrhoea among Akamba', *Tropical and Geographical Medicine*, 31, 139–48.
———— (1991), *Women, Sexuality and the Changing Social Order: the Impact of Government Policies on Reproductive Behaviour in Kenya*, New York: Gordon & Breach.
———— (1994), 'Is there a Distinct African Sexuality? A Critical Response to Caldwell', *Africa*, 64: 2, 220–41.
Alfieri, C. and B. Taverne (2000), 'Ethnophysiologie, des difficultés et complications de l'allaitement maternel chez les Bobo Madare et les Mossi', in A. Desclaux and B. Taverne (eds), *Allaitement et VIH en Afrique de l'Ouest. De l'anthropologie à la santé publique*, Paris: Karthala, 167–87.
Almedom, A. M. (1991a), 'Infant Feeding in Urban Low-income Household in Ethiopia: I, The Weaning Process', *Ecology of Food and Nutrition*, 25, 97–109.
———— (1991b), 'Infant Feeding in Urban Low-income Household in Ethiopia: II, Determinants of Weaning', *Ecology of Food and Nutrition*, 25, 111–21.

Anderson, P. (1983), 'The Reproductive Role of Human Breast', *Current Anthropology*, 24: 1, 25–45.

Ankomath, A. (1998), 'Condom Use in Sexual Exchange Relationships among Young Single Adults in Ghana', *AIDS Education and Prevention*, 10, 303–16.

Ardner, S. (ed.) (1993), *Defining Females. The Nature of Women in Society*, Oxford: Berg Publishers Ltd.

Augé, M. (2000), *Il senso degli altri*, Torino: Bollati-Boringhieri.

Ayodele Adetunji, J. (1991), 'Response of Parents to Five Killer Diseases among Children in Yoruba Community, Nigeria', *Social Science and Medicine*, 12, 1379–87.

Barash, D. (1986), *The Hare and the Tortoise. Culture, Biology, and Human Nature*, New York: Penguin Books.

Barness, L. A. (1993a), 'Nutrizione e problemi nutrizionali. Confronto tra latte materno e latte vaccino', in R. E. Behrman (ed.), *Nelson Trattato di Pediatria*, 14th edition, (Italian edition edited by M. Giovannini), Torino: Medica, 128 ff.

———— (1993b), 'Nutrizione e problemi nutrizionali. Allattamento al seno', in R. E. Behrman (ed.), *Nelson Trattato di Pediatria*, 14th edition, (Italian edition edited by M. Giovannini), Torino: Medica, 124 ff.

Barrel, R. A. and M. G. M. Rowland (1979), 'Infant Food as a Potential Source of Diarrhoeal Illness in Rural West Africa', *Transactions of the Royal Society of Tropical Medicine and Hygiene*, 73: 85.

Barry Lawrance, P. (1994), 'Breast Milk. Best Source of Nutrition for Term and Pre-term Infants', *Care of the Infant. Paediatric Clinics of North America*, 41: 5, 925–41.

Bateson, G . (1980), *Naven*, 2nd edition, London: Wildwood House.

Baylies, C. and J. Bujra (2000), *AIDS, Sexuality and Gender in Africa. Collective Strategies and Struggles in Tanzania and Zambia*, London: Routledge.

Beidelman, T. O. (1960), 'The Baraguyu', *Tanganyika Notes and Records*, 53, 245–78.

———— (1961), 'A Note on Baraguyu House Types and Baraguyu Economy', *Tanganyika Notes and Records*, 56, 56–66.

———— (1982), *Colonial Evangelism: a Socio-Historical Study of an East African Mission at the Grassroots*, Indiana University Press.

Bern, C. et al. (1992), 'The Magnitude of the Global Problem of Diarrhoeal Disease: a Ten-year Update', *Bulletin of the World Health Organization*, 70, 709.

Bledsoe, C. H. (1980), *Women and Marriage in Kpelle Society*, Stanford University Press.

Bleek, W. (1976), 'Spacing of Children, Sexual Abstinence and Breast-feeding in Rural Ghana', *Social Science and Medicine*, 10, 225–30.

Blier, S. (1987), *The Anatomy of Architecture: Ontology and Metaphor in Batammaliba Architectural Expression*, Cambridge: Cambridge University Press.

Blystad, A. (1995), 'Peril or Penalty: AIDS in the Context of Social Change among the Barabaig', in K.-I. Kleep, P. M. Biswalo and A. Talle (eds), *Young People at Risk. Fighting AIDS in Northern Tanzania*, Scandinavian University Press, 86–106.

Bostock, J. (1962), 'Evolutionary Approaches to Infant Care', *Lancet*, 1033.

Bourdieu, P. (1990) [1970], 'The Kabyle House or the World Reversed', in *The Logic of Practice*, Cambridge: Cambridge University Press.

Brandt, A. M. and P. Rozin (eds) (1997), *Morality + Health*, London: Routledge.

Brandtzaeg, B. N. et al. (1981), 'Dietary Bulk as a Limiting Factor for Nutrient Intake in Pre-school Children. III. Studies of Malted Flour from Ragi, Sorghum and Green Gram', *Journal of Tropical Paediatrics*, 27, 184–89.

Broude, G. J. and S. J. Greene (1976), 'Cross-cultural Codes on Twenty Sexual Attitudes and Practices', *Ethnology*, 15: 4, 409–29.

Caillé, A. (1998), *Il terzo paradigma. Antropologia filosofica del dono*, Torino: Bollati-Boringhieri.

———— (2001), 'Note sul paradigma del dono', in G. Maraniello, S. Risaliti, and A. Somaini (eds), *Il dono: Offerta, ospitalità, insidia/The Gift: Generous, Offerings, Threatening, Hospitality*, Milano: Edizioni Charta.

Caldwell, J. C. (1985), 'Strengths and Limitations of the Survey Approach for Measuring and Understanding Fertility Change: Alternative Possibilities', in J. Cleland and J. Hobcraft (eds), *Reproductive Change in Developing Countries*, New York: Oxford University Press.

Caldwell, J. C. and P. Caldwell (1977), 'The Role of Marital Sexual Abstinence in Determining Fertility: A Study of Yoruba in Nigeria', *Population Studies*, 31, 193–217.

———— and ———— (1981), 'The Function of Child-spacing in Traditional Societies and the Direction of Change', in H. J. Page and R. Lesthaeghe (eds), *Child-Spacing in Tropical Africa. Tradition and Change*, New York: Academic Press, 73–91.

———— and ———— (1987), 'The Cultural Context of High Fertility in sub-Saharan Africa', *Population and Development Review*, 13, 409–37.

———— and ———— et al. (1989), 'The Social Context of AIDS in sub-Saharan Africa', *Population and Development Review*, 15, 185–234.

———— and ———— et al. (1991), 'The Destabilization of the Traditional Yoruba Sexual System', *Population and Development Review*, 17, 229–62.

Caldwell, J. C., P. Caldwell and I. O. Orubuloye (1992), 'The Family and Sexual Networking in sub-Saharan Africa: Historical Regional Differences and Present-day Implications', *Population Studies*, 46, 385–410.

Campus, M., M. G. Piredda and M. Mabilia (1998), 'Bibliografia recente in tema dell'allattamento materno', in M. Mabilia (ed.), *PRAE Quaderni del Centro Scientifico Regionale di Prevenzione Sanitaria*, 3, Milano: Aisthesis & Magazine, 205–32.

Caplan, P. (ed.) (1993), *The cultural Construction of Sexuality*. London: Routledge.

Carnell, W. J. (1955a), 'Sympathetic Magic among the Gogo of Mpwapwa District', *Tanganyika Notes and Records*, 39, 25–38.

———— (1955b), 'Four Gogo Folk Tales', *Tanganyika Notes and Records*, 40, 30–42.

Carsten, J and S. Hugh-Jones (eds) (1995), *About the House. Lévy Strauss and Beyond*, Cambridge: Cambridge University Press.

Cixous, H. (1997), 'Sorties: Out and Out: Attacks/Ways Out/Forays', in A. D. Schrift (ed.), *The Logic of the Gift. Towards an Ethic of Generosity*, London: Routledge, 148–73.

Cluass, H. (1911), 'Die Wagogo', *Baessler Archiv*, 2, 1–72. Berlin: Reimer.

Cohen, M. N. (1989), *Health and the Rise of Civilization*, New Haven: Yale University Press.

Cole, H. (1902), 'Notes on the Wagogo of German East Africa', *Journal of Anthropological Institute*, 32, 305–38.

Cordell, O. T. (1935), *Gogo Grammar*, Dar es Salaam.

Creyton, M. L. (1992), 'Breastfeeding and Baraka in Northern Tunisia', in V. Maher (ed.), *The Anthropology of Breastfeeding*, Oxford: Berg Publishers Limited, 37–58.

Cunningham, A. S. (1995), 'Breastfeeding: Adaptive Behaviour for Child Health and Longevity', in P. Stuart-Macadam and K. A. Dettwyler (eds), *Breastfeeding. Biological Perspectives*, New York: Aldine De Gruyter, 243–64.

Cutting, A. et al. (1981), 'A World-wide Survey on the Treatment of Diarrhoeal Disease by Oral Rehydration in 1979', *Annual Tropical Paediatric*, 1, 199–208.

Davidson, J. and D. Meyers (1992), *No Time To Waste, Poverty and the Global Environment*, United Kingdom and Ireland: Oxfam.

Dean, N. R. (1994), 'A Community Study of Child Spacing, Fertility and Contraception in West Pokot District, Kenya', *Social Science and Medicine*, 38, 1575–84.

De Beauvoir, S. (1961), *Il secondo sesso*, Milano: Il Saggiatore.

De Bruyn, M. (1992), 'Women and AIDS in Developing Countries', *Social Science and Medicine*, 34, 249–62.

De Giacomo, C. and G. Maggiore (1999), *Gastroentologia, epatologia e nutrizione. Un approccio ai problemi*, Torino: UTET.

De Loache, J. and A. Gottlieb (eds) (2000), *A world of Babies. Imagined Childcare Guides for Seven Societies*. Cambridge: Cambridge University Press.

De Sousa, J. F. (1991), 'Traditional Beliefs and Practices Related to Childhood Diarrhoeal Disease in a High-Density Suburb of Maputo', B. A. thesis, Department of Sociology, University of Zimbabwe.

Dettwyler, K. A. (1986), 'Infant Feeding in Mali, West Africa: Variations in Belief and Practice', *Social Science and Medicine*, 23: 7, 651–64.

———— (1987), 'Breastfeeding and Weaning in Mali: Cultural Context and Hard Data', *Social Sciences and Medicine*, 24: 8, 633–44.

———— (1988), 'More than Nutrition: Breastfeeding in Urban Mali', *Medical Anthropological Quarterly*, 2, 172–83.

De Zoysa, I. et al. (1984), 'Perceptions of Childhood Diarrhoea and its Treatment in Rural Zimbabwe', *Social Science and Medicine*, 19, 727–34.

Dixon Whitaker, E. (1994), 'L'intersezione di processi storici e bioculturali sull'"organismo materno". Pratiche nutritive infantili nel periodo fascista', in A. Destro (ed.), *Le politiche del corpo. Prospettive antropologiche e storiche*, Bologna: Patron Editore, 131–80.

———— (2000), *Measuring Mamma's Milk. Fascism and the Medicalization of Maternity in Italy*, Ann Arbor: University of Michigan Press.

Douglas, M. (1989), 'Il n'y a pas de don gratuit. Introduction à la traduction anglaise de l'Essai sur le don, *Revue du Mauss*, 4, 99–115.

Dow, T. E. (1977), 'Breastfeeding and Abstinence among the Yoruba', *Studies in Family Planning*, 8, 208–18.

Durham, W. (1991), *Coevolution. Genes, Culture, and Human Diversity*, Stanford: Stanford University Press.

Evans-Pritchard, E. E. (1940), *The Nuer: a Description of the Modes of Livelihood and Political Institutions of a Nilotic People*. Oxford: Oxford University Press.

———— (1953), 'Nuer Spear Symbolism', *Anthropological Quarterly*, 26, 1–19.

———— (1956), *Nuer Religion*, Oxford: Oxford University Press.

Fildes V. A. (1986), *Breasts, Bottles and Babies*. Edinburgh: Edinburgh University Press.

———— (1995), 'The Culture and Biology of Breastfeeding: an Historical Review of Western Europe', in P. Stuart-Macadam and K. A. Dettwyler (eds), *Breastfeeding Biocultural Perspectives*, New York: Aldine De Gruyter, 101–126.

———— (1997), *Madre di latte. Balie e baliatico dall'antichità al XX secolo*, Milano: Edizioni San Paolo, (Originally published as *Wet Nursing. A History from antiquity to the Present*, Oxford: Basil Blackwell, 1988).

Fishbein, M. and F. Ajzen (1975), *Belief, Attitude, Intention and Behaviour: an Introduction to Theory and Research*, Reading, MA: Addison Wesley.

Fomon, F. J. (1974), *Infant Nutrition*, 2nd edition, Philadelphia: W.B. Saunders, 360 ff.

Geertz, Clifford (1974), 'From the Native's Point of View: on the Nature of Anthropological Understanding', *Bulletin of American Academy of Arts and Sciences*, 28: 1, (Italian translation 1988, in *Antropologia interpretativa*, Bologna: Il Mulino, 71–90).

———— (1987), 'Verso una teoria interpretativa della cultura', in *Interpretazioni di culture*, Bologna: Il Mulino, 39–72.

———— (1988), 'Il senso comune come sistema culturale', in *Antropologia interpretativa*, Bologna: Il Mulino, 91–117.

Glasier, A. and A. S. McNeilly (1990), 'Anatomy and Development of the Breast', in *Bailliere's Clinical Endocrinology and Metabolism*, vol. 4, 379–95.

Glen Nakano, E., Chang, G. and L. Rennie Forcey (eds) (1994), *Mothering. Ideology, Experience, and Agency*, London: Routledge.

Gluckman, M. (1950), 'Kinship and Marriage among the Lozi of Northern Rhodesia and Zulu of Natal', in A. R. Radcliffe-Brown and D. Forde (eds), *African Systems of Kinship and Marriage*, London: Oxford University Press for the International African Institute, eleventh impression 1975, 166–206.

Godbout, J. T. (1993), *Lo spirito del dono*, Torino: Bollati-Boringhieri.

————— (1998), *Il linguaggio del dono*,Torino: Bollati-Boringhieri.

Göksen, F. (2002), 'Normative vs Attitudinal Considerations in Breast-feeding Behaviour: Multifaceted Social Influences in a Developing Country Context', in *Social Science and Medicine* 54, 1743–53.

Goodman, A. H. and T. L. Leatherman (eds) (1998), *Building a New Biocultural Synthesis*, Ann Arbor: University of Michigan Press.

Goody, J. (1973), 'Bridewealth and Dowry in Africa and Eurasia', in J. Goody and S. J. Tambiah, *Bridewealth and Dowry*, Cambridge: Cambridge University Press.

Gottlieb, A. (2000), 'Luring Your Child into this Life: A Beng Path for Infant Care', in J. De Loache and A. Gottlieb (eds) (2000), *A World of Babies. Imagined Childcare Guides for Seven Societies*, Cambridge: Cambridge University Press, 55–90.

Green, E. C. (1985), 'Traditional Healers, Mothers and Childhood Diarrhoeal Disease in Swaziland: the Interface of Anthropology and Health Education', in *Social Science and Medicine*, 20, 277–85.

Green, E. C. et al. (1994), 'The Snake in the Stomach: Child Diarrhoea in Central Mozambique', *Medical Anthropology Quaterly*, 8, 4–24.

Gulliver, P. H. (1955), *The Family Herds: a Study of Two Pastoral Tribes in East Africa, the Jie and the Turkana*, London: Routledge.

————— (1959), 'A Tribe Map of Tanganyika', in *Tanganyika Notes and Records*, 52, 61–74.

Gunnlaugsson, G. and J. Einarsdóttir (1993), 'Colostrum and Ideas about Bad Milk: a Case Study from Guinea-Bissau', in *Social Science and Medicine*, 36: 3, 283–88.

Hadjivayanis G. G. (1989), 'La main mise sur la paysannerie gogo', in *Urafiki*, 43, 2–10.

Hall E. (1966), *The Hidden Dimension*, New York: Doubleday & Co. Inc.

Haram, L. (1995), 'Negotiating Sexuality in Times of Economic Want: the Young and Modern Meru Women', in K. I. Klepp, P. M. Biswalo and A. Talle (eds), *Young People at Risk. Fighting AIDS in Northern Tanzania*, Scandinavian University Press.

————— (1999), *'Women out of Sight': Modern Women in Gendered Worlds. The Case of Meeru of Northern Tanzania*. University of Bergen.

————— (2001), '"In Sexual Life Women are Hunters": AIDS and Women who Drain Men's Body. The Case of the Meru of Northern Tanzania', *Society in Transition*, 32, 47–55.

————— (2003), 'Prostitutes or Modern Women? Negotiating Sexuality in Northern Tanzania', Forthcoming in Signe Arnfred (ed.), *Rethinking Sexuality*, Uppsala: The Nordic Africa Institute.

Hartnoll, A. V. (1932), 'The Gogo Mtemi: Information Collected from the Two Native Councils Dodoma North and Dodoma South, Tanganyika Territory', *South African Journal of Science*, 29, 734–41.

Hassig, S. E. et al. (1991), 'Duration and Correlates of Post Partum Abstinence in Four Sites in Zaire', *Social Science and Medicine*, 32, 343–47.

Heald, S. (1995), 'The Power of Sex: Some Reflection on the Caldwells' "African Sexuality" Thesis', *Africa*, 65: 4, 489–505.

Helman, C. G. (1990), *Culture, Health and Illness*, Oxford: Butterworth-Heinemann Ltd.

Helsing, E. and F. Savage King (1983), *Breastfeeding in Practice*, Oxford: Oxford University Press.

Hogel, J. et al. (1991), *Indigenous Knowledge and Management of Childhood Diarrhoea Diseases*, Arlington, VA: Management Sciences for Health-PRITECH Project.

Hooton, J. W. L. et al. (1991), 'Human Colostrum Contains an Activity that Inhibits the Production of IL-2', *Clinical Experimental Immunology*, 86, 520–24.

Howard, M. (1994), 'Socio-economic Causes and Cultural Explanation of Childhood Malnutrition among the Chagga of Tanzania', *Social Science and Medicine*, 38, 239–51.

Huygens, P. et al. (1996), 'Rethinking Methods for the Study of Sexual Behaviour', *Social Science and Medicine*, 42: 2, 221–31.

Iliffe, J. (1979), *A Modern History of Tanganyika*, Cambridge: Cambridge University Press.

Isenalumhe, A. E. and O. Oviawe (1986), 'The Changing Pattern of Post Partum Sexual Abstinence in Nigerian Rural Community', *Social Science and Medicine*, 23, 683–86.

Iyun, B. F. (1992), 'Women's Status and Childhood Mortality in Two Areas in South-Western Nigeria: a Preliminary Analysis', *GeoJournal*, 26: 1, 43–52.

———— (2000), 'Environment Factors, Situation of Women and Child Mortality in South-Western Nigeria', *Social Science and Medicine*, 51: 10, 1473–89.

Janzen, J. (1978), *The Quest for Therapy in Lower Zaire*, Berkeley: University of California Press.

———— (1987), 'Therapy. Management: Concept, Reality, Process', *Medical Anthropology Quarterly*, 1, 68–84.

———— (1989), 'Health, Religion, and Medicine in Central and Southern African Traditions', in L. E. Sullivan (ed.), *Health and Restoring. Health and Medicine in the World's Religious Traditions*, New York: MacMillan Publishing Company, 225–54.

Jelliffe, D. B. and J. Bennett (1972), 'Aspects of Child Rearing in Africa', *J. Trop. Paediat. Environ. Child Health*, 18, 26–43.

Jelliffe, D. B. and E. F. Jelliffe (1978), *Human Milk in the Modern World: Psychology, Nutritional and Economic Significance*. Oxford: Oxford University Press.

———— (1980), 'Feeding in Early Infancy and Primary Health Care', *Ecology of Food and Nutrition*, 9, 189–94.

———— (1986), The uniqueness of Human Milk Up-date: Ranges of Evidence and Emphases in Interpretation' in D. B. Jelliffe and E. F. Jelliffe (eds), *Advances in International Maternal and child Health*, vol. 6, Oxford: Clarendon Press, 129–47.

Johnson, M. C. (2000), 'The View from the Wuro: A Guide to Child Rearing', in J. De Loache and A. Gottlieb (eds), *A World of Babies. Imagined*

Childcare Guides for Seven Societies, Cambridge: Cambridge University Press, 171–98.

Johnston, H. H. (1919), 'The Usagara-Ugogo Languages, Group G', in *A Comparative Study of the Bantu and Semi-Bantu Languages*, Oxford: Clarendon Press, 141–53.

Kavishe, F. P. and O. Yambi (1991), 'Malnutrition', in G. M. P. Mwaluko et al., *Health & Disease in Tanzania*, London: Harper Collins Academic.

Kenya. P. R. et al. (1990), 'Oral Rehydration Therapy and Social Marketing in Rural Kenya', *Social Science and Medicine*, 31, 979–87.

Kenyatta, J. (1938), *Facing Mount Kenya. The Tribal Life on the Gikuyu*, London: Secker & Warburg.

Kesby, M. (2000), 'Participatory Diagramming as a Means to Improve Communication about Sex in Rural Zimbabwe: a Pilot Study', *Social Science and Medicine*, 50, 1723–41.

Killewo, J. et al. (1994a), 'Prevalence of HIV-1 in the Kagera Region of Tanzania: a Population Based Study', in J. Killewo (ed.), *Epidemiology towards the Control of HIV Infection in Tanzania with Special Reference to the Kagera Region*, Sweden: UMEA, Part I, 1081–85.

———— (1994b), 'Socio-geographical Patterns of HIV-1 Transmission in Kagera Region, Tanzania', *Social Science and Medicine*, 38, 129–34.

———— (1994c), 'Knowledge, Attitudes and Perceptions Regarding HIV Infection Risk. Communicating with the People as a Basis for Planning Interventions', in J. Killewo (ed.), *Epidemiology towards the Control of HIV Infection in Tanzania with Special Reference to the Kagera Region*, Sweden: UMEA, Part IV, 1–16.

———— (1994d), 'Incedence of HIV-1 Infection among Adults in the Kagera Region of Tanzania', in J. Killewo (ed.), *Epidemiology towards the Control of HIV Infection in Tanzania with Special Reference to the Kagera Region*, Sweden: UMEA, Part V, 528–36. (First published in *International Journal of Epidemiology*, 22: 3, 1993.)

Kitzinger, S. (1978), *Women as Mothers: How They See Themselves in Different Cultures*, New York: Vintage Books.

———— (1980), *The Experience of Breastfeeding*, London: Penguin.

———— (1994), *Ourselves as Mothers*, New York: Addison Wesley.

Klepp, K.-I., P. M. Biswalo and A. Talle (eds) (1995), *Young People at Risk. Fighting AIDS in Northern Tanzania*, Scandinavian University Press.

Knutsson, K. E. and T. Melbin (1969), 'Breastfeeding Habits and Cultural Context', *Journal of Tropical Pediatrics*, 15, 40–49.

Konner, M. (1982), *The Tangled Wing. Biological Constraints on the Human Spirit*, New York: Holt, Reinhart and Winston.

Kwebena Nketia, J. H. (1968), 'Multi-part Organization in the Music of the Gogo of Tanzania', in *Tanzania Notes and Records*, 68, 63–73.

Kwesigabo, G. et al. (1998), 'Decline in the Prevalence of HIV-1 in Young Women in the Kagera Region of Tanzania', *Journal of Acquired Deficiency Syndromes and Human Retrovirology*, 17, 262–68.

———— (2001), *Trend of HIV Infection in the Kagera Region of Tanzania 1987–2000*, Sweden: UMEA University.

La Fontaine, J. S. (1972), 'Ritualisation of Women's Life Crises in Bugisu', in J. S. La Fontaine (ed.), *The Interpretation of ritual*, London: Tavistock Publications.

———— (1977), 'The Power of Rights', *Man* (N.S.), 12, 421–37.

(1981), 'The Domestication of the Savage Male', *Man* (N.S.), 16, 333–49.

Last, M. (1993), 'Non-western Concepts of Disease', in W. F. Bynum and R. Porter (eds), *Companion Encyclopedia of the History of Medicine*, London: Routledge, 634–59.

Lawrence, R. A. (1994), *Breastfeeding: a Guide for the Medical Profession*, 4th edition, St. Louis: C.V. Mosby.

Leach, V. (1990), *Women and Children in Tanzania. A situation Analysis*, Dar es Salaam: Government of the United Republic of Tanzania and United Nations Children's Fund.

Leavitt, S. C. (1991), 'Sexual Ideology and Experience in a Papua New Guinea Society', *Social Science and Medicine*, 33: 8, 895–907.

Le Blanc, M-N. et al. (1991), 'The African Sexual System: Comment on Caldwell et al.', *Population and Development Review*, 17: 3, 497–505.

Lesthaeghe, R. et al. (1981), 'Child-spacing and Fertility in sub-Saharan Africa: an Overview of Issues', in H. J. Page and R. Lestaeghe (eds), *Child-Spacing in Tropical Africa. Tradition and Change*, New York: Academic Press, 3–23.

Le Vine, R. (1964), 'The Gusy Family', in R. F. Gray and P. H. Gulliver (eds), *The Family Estate in Africa*, Boston, MA: Boston University Press, 63–82.

Lind, J., V. Vuorenkoski and O. Wasz-Hockert (1971), *Psychosomatic Medicine in Obstetrics and Gynaecology*, London: Third International Congress.

Lindebaum, S. (1991), 'Anthropology Rediscovers Sex. Introduction', *Social Science and Medicine*, 33: 8, 865–66.

Loizos, P. and P. Heady (eds) (1999), *Conceiving Persons. Ethnographies of Procreation, Fertility and Growth*, London: The Athlone Press.

Lwihula, G. et al. (1994), 'AIDS Epidemic in Kagera Region, Tanzania, the Experience of Local People', in J. Killewo (ed.), *Epidemiology towards the Control of HIV Infection in Tanzania with Special Reference to the Kagera Region*, Sweden: UMEA, 347–57. (First published in AIDS Care, 5: 3, 1993.)

Mabilia, M. (1991), *Il valore sociale del cibo*, Milano: Angeli.

———— (1995), 'Risks to Breastfeeding: Beliefs, Attitudes and Behaviour in Infant Breastfeeding among the Gogo Mother. (Dodoma Rural District, Tanzania)', Lecture to Centre for Cross-Cultural Research on Women, Queen Elizabeth House, Oxford University.

———— (1996a), 'Beliefs and Practices in the Breastfeeding and Weaning among the Wagogo of Chigongwe, Dodoma Rural District, Tanzania. I. Breastfeeding', *Ecology of Food and Nutrition*, 35, 195–207.

———— (1996b), 'Beliefs and Practices in the Breastfeeding and Weaning among the Wagogo of Chigongwe, Dodoma Rural District, Tanzania. II Weaning', *Ecology of Food and Nutrition*, 35, 209–17.

————— (1998), 'Allattamento: processo composito e complesso. Un caso etnografico', in M. Mabilia (ed.), *Allattamento materno*, PRAE 3, Quaderni del Centro Scientifico Regionale di Prevenzione Sanitaria, Cagliari,Casa Editrice Aisthesis, Milano, 101–16.

————— (1999), 'Risks to Breastfeeding: Baby's Health and Mothers' Behaviour: Interactions', Lecture to Department of Biological Anthropology, Duke University.

————— (2000), 'The Cultural Context of Childhood Diarrhoea among Gogo Infants', *Anthropology and Medicine*, 7: 2, 191–208.

————— (2001a), 'Allattamento come dono? Un caso etnografico', *DiPAV*, 2, 115–36.

————— (2001b), 'Alcune considerazioni da una prospettiva antropologica sulla diffusione dell'HIV/AIDS nell'Africa Sub-Sahariana', *Salute e sviluppo* (N.S.) 2, 29–35. (Special English edition, 'Some Anthropological Perspectives on the Spread of HIV/AIDS in the Sub-saharan Africa', *Health and Development*, 2, 29–36.)

MacCormack, C. P. (ed.) (1982), *Ethnography of Fertility and Birth*, London: Academic Press.

Macy, I. G., H. J. Kelly and S. Re (1953), 'Decomposition of Milks', in NAS-NRC Publ., 254.

Maddox, G. H. (1990), '*Mtunya*: Famine in Central Tanzania, 1917–1920', *Journal of African History*, 31: 2, 181–98.

————— (1995), 'Introduction: The Ironies of *History, Mila na Desturi za Wagogo*', in M. E. Mnyampala, *The Gogo History, Customs, and Traditions*, translated, introduced and edited by G. H. Maddox. London: M.E. Sharpe.

Maher, V. (1992), 'Breast Feeding and Maternal Depletion: Natural Law or Cultural Arrangements?', in V. Maher (ed.), *The anthropology of Breast-feeding*, Oxford: Berg Publishers Limited.

————— (ed.) (1992), *The anthropology of Breast-feeding*, Oxford: Berg Publishers Limited.

————— (1992) (ed.), *Il latte materno. I condizionamenti culturali di un comportamento*, Torino: Rosenberg & Sellier).

Malinowski, B. (1967), *A Diary in the Strict Sense of the Term*, London: The Athlone Press.

Maraniello, G., S. Risaliti and A. Somaini (eds) (2001), *Il dono/The Gift. Offerta, ospitalità, insidia/Generous, Offerings, Threatening, Hospitality*, Milano: Edizioni Charta.

Mauss, M. (1965), 'Il saggio sul dono. Forma e motivo dello scambio nelle società arcaiche', in Marcel Mauss, *Teoria generale della magia e altri saggi*, Torino: Einaudi, 153–229.

McCormack, C. O. (1988), *Ethnology of Fertility and Birth*, London: Academic Press.

MCH (1989/90), *Mother and Child Health Report*. Dodoma Hospital. Unpublished Annual Record.

McKenna, J. J. and S. Mosko (1993), 'Evolution and Infant Sleep: an Experimental Study of Infant-Parent Co-sleeping and its Implications for SIDS', *Acta Paediatrica Supplement*, 389, 31–6.

Merchant, C. (1988), *La morte della natura*, Milano: Garzanti.

Mnyampala, M. E. (1954), *Historia Mila na Desturi za Wagogo*, Dar es Salaam: East African Literature Bureau.

Mohrbacher, N. and J. Stock (eds) (1997), *The Breastfeeding Answer Book*, Illinois: La Leche League International, Schaumburg.

Moller, M. S. G. (1961), 'Custom, Pregnancy and Child Rearing in Tanganyika', *African Child Health*, 7, 66–80.

Moore, H. L., T. Sanders and B. Kaare (eds) (1999), *Gender, Fertility and Transformation in East and Southern Africa*, London: The Athlone Press.

Mosha, A. C. and U. Svanberg (1990), 'The Acceptance and Intake of Bulk-reduced Weaning Food: the Luganga Village Study', *Food Nutrition Bulletin*, 20, 69–74.

Murdock, G. P. (1967), 'Ethnographic Atlas: a Summary', *Ethnology*, 6, 109–236.

Mutanda, L. N. (1980), 'Epidemiology of Acute Gastroenteritis in Early Childhood in Kenya', *East African Medical Journal*, 57, 545.

Mwaluko, G. M. P. et al. (1991), *Health and Disease in Tanzania*, London: Harper Collins Academic.

Needam, R. (1960), 'The Left Hand of the Mugwe: an Analytical Note on the Structure of Meru Symbols', *Africa*, 30, 20–33.

———— (ed.) (1973), *Left and Right*, Illinois: University of Chicago Press.

Nyerere, J. (1968), *Ujamaa. Essays on Socialism*, Oxford: Oxford University Press.

———— (1969), *Freedom and Socialism*, Dar es Salaam.

Oakley, A. (1985), *Sex, Gender, and Society*. Redwood Books.

Ojofeitimi, E. O. (1981), 'Mothers' Awareness of Benefits of Breast Milk and Cultural Taboos during Lactation', *Social Science and Medicine*, 15E, 135–38.

Okediji, O. F. et al. (1976), 'The Changing African Family Project: a Report with Special Reference to the Nigerian Segment', *Studies in Family Planning*, 7, 126–36.

Oni, G. A. (1987), 'Breastfeeding: its Relationship with Postpartum Amenorrhea and Postpartum Sexual Abstinence in a Nigerian Community', *Social Science and Medicine*, 24: 3, 255–62.

Orubuloye, I. O. (1979), 'Sexual Abstinence Patterns in Rural Western Nigeria: Evidence from Survey of Yoruba Women', *Social Science and Medicine*, 13A, 667–72.

Osservatorio Italiano Sulla Salute Globale (2004), *Rapporto 2004 Salute e Globalizzazione*, Milano: Feltrinelli.

Page, H. J. and R. Lestaeghe (eds) (1981), *Child-Spacing in Tropical Africa. Tradition and Change*, New York: Academic Press.

Palmer, G. (1990), *The Politics of Breastfeeding*, London: Pandora Press.

Patel, V. P. (1988), 'Casual Reasoning and the Treatment of Diarrhoeal Disease by Mothers in Kenya', *Social Science and Medicine*, 27, 1277–86.

Piccone Stella, S. and C. Saraceno (1996), *Genere. La costruzione sociale del femminile e del maschile*, Bologna: Il Mulino.

Pomata, G. (1983), 'La storia delle donne: una questione di confine', in *Il mondo contemporaneo. Gli strumenti della ricerca 2. Questioni di metodo*, La Nuova Italia, 1439–69.

Quandt, S. A. (1985), 'Biological and Behavioural Predictors of Exclusive Breastfeeding Duration', *Medical Anthropology*, 9, 139–151.

———— (1986), 'Patterns of Variation in Breastfeeding Behaviours', *Social Science and Medicine*, 23, 445–453.

———— (1995), 'Sociocultural Aspects of the Lactation Process', in P. Stuart-Macadam and K. A. Dettwyler (eds), *Breastfeeding. Biocultural Perspectives*, New York: Aldine De Gruyter, 127–43.

Quinn, T. C. et al. (1986), 'AIDS in Africa: an Epidemiology Paradigm', *Science*, 234: 4779, 955–63.

Radcliffe-Brown, A. and D. Forde (eds) (1950), *African Systems of Kinship and Marriage*, Oxford: International African Institute, Oxford University Press.

Raphael, D. (1966), 'The Lactation-Suckling Process in the Matrix of Supportive Behaviour', Ph.D. thesis, Colombia University Microfilm # 69–15,580.

———— (1981), 'The Midwife as Doula: A Guide to Mothering the Mother', *Journal of Nurse-Midwifery*, November-December, 13–15.

———— (1984), 'Weaning is always: the Anthropology of Breast-feeding Behaviour', *Ecology of Food and Nutrition*,15, 203–13.

———— (ed.) (1985), *Only Mothers Know. Patterns of Infant Feeding in Traditional Culture*, Westport, CT: Greenwood Press.

Raphael, D. and F. Davis (1985), *Patterns of Infant Feeding in Traditional Culture*, Westport, CT: Greenwood Press.

Rich, A. (1995), *Of Woman Born. Motherhood as Experience and Institution*, W.W. Norton & Company Ltd.

Richards, A. (1939), *Land, Labour and Diet in northern Rhodesia*, Oxford: International African Institute, Oxford University Press.

———— (1956), *Chisungu. A Girl's Initiation Ceremony among the Bemba of Zambia*, London: Tavistock Publications.

Riordan, J. (1993), 'The Cultural Context of Breastfeeding', in J. Riordan and K. Auerbach (eds), *Breastfeeding and Human Lactation*, Boston: Jones and Bartlett, 27–48.

Riordan, J. and K. Auerbach (eds) (1993), *Breastfeeding and Human Lactation*, Boston and London: Jones and Bartlett.

Rubin, G. (1975), 'The Traffic in Women: Notes on the "Political Economy" of Sex', in R. Reiter (ed.), *Towards an Anthropology of Women*, New York, Monthly Review Press, 157–210.

Saucier, J. F. (1972), 'Correlates of the Long Post Partum Taboo: a Cross-cultural Study', *Current Anthropology*, 13, 238–49.

Savage King, F. and A. Burgess (1993), *Nutrition for Developing Countries*, Oxford: Oxford University Press.

Schaegelen, R. P. T. (1938a), 'La tribu des Wagogo', *Anthropos*, 33: 2, 195–217.

——— (1938b), 'La tribu des Wagogo', *Anthropos* 33: 2, 515–67.

Scheper-Hughes, N. (1984), 'Infant Mortality and Infant Care Economic and Cultural Constraints on Nurturing in Northeast Brazil', *Social Science and Medicine*, 19, 535-46.

——— (1987a), 'The Mindful Body: a Prolegomenon to Future Work in Medical Anthropology, *Medical Anthropology Quarterly*, 1: 1, 6–41.

——— (1987b), 'The Cultural Politics of Child Survival', in N. Scheper-Hughes (ed.), *Child Survival: Anthropological Perspectives on the Treatment and Maltreatment of Children*, Dordrecht: Reidel, 1–29.

——— (1987c), 'Culture, Scarcity and Maternal Thinking: Mother Love and Child Death in Northeast Brazil', in N. Scheper-Hughes (ed.), *Child Survival: Anthropological Perspectives on the Treatment and Maltreatment of Children*, Dordrecht: Reidel, 187–208.

——— (1988), 'Letter to the editor', *Culture Medicine and Psychiatry*, 12, 259–60.

——— (1992), *Death Without Weeping: The Violence of Everyday Life in Brazil*, Berkeley, CA: University of California Press.

Scheper-Hughes, N. and M. M. Lock (1987), 'The Mindful Body: a Prolegomenon to Future Work in Medical Anthropology, *Medical Anthropology Quarterly*, 1: 1, 6–41.

Schoenmaekers, R. et al. (1981), 'The Child-spacing Tradition and the Postpartum Taboo in Tropical Africa: Anthropological Evidence', in H. J. Page and R. Lesthaeghe (eds), *Child-Spacing in Tropical Africa. Tradition and Change*, New York: Academic Press, 25–71.

Segalen, V. (1982), *Le parole perdute*, Milano: Jaca Book. (Originally *Les Immémoriaux*, 1907).

Sen, A. (1990), 'Gender and Cooperative Conflicts', in I. Tinker, *Persistent Inequalities*, Oxford: Oxford University Press, 123–49.

Serventi, M., (1991), 'Malnutrizione infantile. Un problema di tutti (e quindi di nessuno)', *CUAMM Notizie, Salute e Sviluppo*, IV, 34–6.

Shapera, I. (1971), *Married Life in an African Tribe*, Harmondsworth: Penguin Books.

Silberschmidt, M (1999), *"Women Forget that Men are the Masters". Gender Antagonism and Socio-Economic Change in Kiisi District, Kenya*, Stockholm: Elanders Gotab.

Smith Oboler, R. (1994), 'The House-property Complex and African Social Organization', *Africa*, 64: 3, 342–58.

Southon, E. J. (1881), 'Notes of the Journey through Northern Ugogo, in East Central Africa, in July and August 1879', *Proceedings of the Royal Geographical Society*, III, 547–53.

Steinberg, S. (1996), 'Childbearing Research: a Trans-cultural Review', *Social Science and Medicine*, 43: 12, 1765–84.

Strathern, M. (1987), *Dealing with inequality. Analysing Gender Relations in Melanesia and Beyond*, Cambridge: Cambridge University Press.

Stuart-Macadam, P. (1995), 'Biocultural Perspectives on Breastfeeding', in P. Stuart-Macadam and K. A. Dettwyler (eds) (1995), *Breastfeeding Biocultural Perspectives*, New York: Hawthorne, 1–38.

Stuart-Macadam, P. and K. A. Dettwyler (eds) (1995), *Breastfeeding Biocultural Perspectives*, New York: Hawthorne.

Talle, A. et al. (1995), 'Introduction', in K-I Klepp, P. M. Biswalo and A. Talle (eds), *Young People at Risk. Fighting AIDS in Northern Tanzania*, Scandinavian University Press, 17–18.

Taylor, M. (1985), *Transcultural Aspects of Breastfeeding*, La Leche League International's Lactation Consultant Series. Unit 2. Wayne, NJ: Avery Publication.

Thiele, G. (1984a), 'Location and Enterprise Choice: a Tanzanian Case Study', *Journal of Agricultural Economics*, 35: 2, 257–64.

———— (1984b), 'State Intervention and Commodity Production in Ugogo: a Historical Perspective', *Africa*, 54: 3, 92–107.

———— (1985), 'Villages as Economic Agents: The Accident of Social Reproduction', in R. G. Abrahams (ed.), Villagers, Villages and the State in Modern Tanzania. Cambridge African Monograph 4, 81–109.

———— (1986) 'The Tanzanian Villagisation Programme: Its Impact on Household Production in Dodoma', *Canadian Journal of African Studies*, 243–58.

Thompson, J. A. (1996), 'A Biocultural Approach to Breastfeeding', *New Beginnings*, 13: 6, 164–67.

Tremayne, S. (ed.) (2001), *Managing Reproductive Life. Cross-Cultural Themes in Fertility and Sexuality*, Oxford: Berghahn Books.

Turner, V. (1967), *The Forest of Symbols. Aspects of Ndembu Ritual*, Ithaca: Cornell University Press.

Tuzin, D. (1991), 'Sex, Culture and the Anthropologist', *Socioal Science and Medicine*, 33: 8, 867–74.

UNICEF (1991), *L'allattamento al seno. Protezione, incoraggiamento e sostegno.* Collana di Pedagogia Sanitaria, UNICEF: Comitato italiano.

———— (1999), *Breastfeeding: Foundation for a Healthy Future*, UNICEF: Goals 2000.

———— (2002), *La salute è vita*, (terza edizione), New York: UNICEF.

———— (2003), *La condizione dell'infanzia nel mondo*. Rapporto UNICEF 2003, Roma: Comitato Italiano per l'UNICEF.

Vance, C. S. (1991), 'Anthropology Rediscovers Sexuality: a Theoretical Comment', *Social Science and Medicine*, 33: 8, 875–84.

Walker, A. R. P. and F. I. Adam (2000), 'Breastfeeding in sub-Saharan Africa: Outlook 2000', *Public Health and Nutrition*, 3: 3.

WHO (1985), *The Quantity and Quality of Breast Milk: Report on the WHO Collaborative Study of Breastfeeding*, Geneva: World Health Organization.

———— (1994), 'Women, Sex and AIDS', in World Health Organization (ed.), *AIDS: Images of the Epidemic*, Geneva: World Health Organization, 56–61.

———— (2002), *The World Health Report 2002: Reducing Risks, Promoting Healthy Life*, Geneva: WHO.

WHO-UNICEF-USAID (1992), 'Indicators for Assessing Breastfeeding Practices', *Update*, No. 10, February.

Williams, C. D. (1935), 'Kwashiorkor: Nutrition Diseases of Children Associated with a Maize Diet', *Lancet*, 2, 1151–52.

Wilson, E. O. (1978), *On Human Nature*, Cambridge MA: Harvard University Press.

Woodruff, A. W. (ed.) (1974), *Medicine in the Tropics*, The English Language Book Society and Churchill Livingstone.

Woolridge, M. W. (1995), 'Baby-controlled Breastfeeding: Biocultural Implications', in P. Stuart-Macadam and K. A. Dettwyler (eds), *Breastfeeding. Biocultural Perspectives*, New York: Aldine De Gruyter, 217–42.

World Development Report (1990), Oxford: Oxford University Press.

Yoder, P. S. (1981), 'Knowledge of Illness and Medicine among Cokwe of Zaire', *Social Science and Medicine*, 15B, 237–45.

———— (1995), 'Examining Ethnomedical Diagnosis and Treatment Choices for Diarrhoeal Disorders in Lumumbashi Swahili', *Social Science and Medicine*, 16, 211–47.

INDEX

48, 52, 54, 53, 56, 57, 63nn. 15, 16, 17, 64n. 26, 85, 86, 89, 101, 109, 114; production of, 47, 56; *real*, 45, 54, 64n. 25; sour, 57, 58
miti, 51, 64n. 29, 77, 92n. 2, 94n. 38
miti ye cigogo, 77, 89, 90
mixed feeding, 2, 20, 48–51, 78, 80
mixture, 84, 87, 89, 107, 110, 111, 112
mother/s, 1, 2, 3, 4, 5, 10, 13, 14, 15, 17, 20, 21, 22, 26, 27, 28, 29, 30, 33, 34, 36, 37, 41, 42, 43, 44, 45, 46, 47, 48, 49, 50, 51, 52, 53, 54, 55, 56, 57, 58, 59, 60, 61, 61n. 5, 62nn. 8, 12, 13, 63nn. 14, 15, 16, 17, 22, 64nn. 26, 28, 66, 68, 69, 70, 71, 72, 73, 74, 75, 76, 78, 79, 80, 81, 82, 83, 84, 85, 86, 87, 88, 89, 90, 91, 91n. 1, 92nn. 4, 9, 11, 12, 93n. 21, 94n. 37, 95n. 46, 97n. 65, 98nn. 70, 75, 76, 99nn. 82, 85, 100, 101, 102, 103, 104, 105, 106, 107, 108, 109, 110, 111, 112, 113, 114, 115, 115n. 3, 116nn. 5, 6, 118n. 28; bad, 66, 88, 105, 108, 109, 113, 115; good, 37, 66, 67, 68, 82, 83, 100, 102–105, 108, 112, 115; *see also* behaviour, dyad, milk
motherhood, 20, 37, 68–72
mothering, 37, 68–72
mother-in-law, 36, 37, 44, 70, 75, 82, 91, 104, 111
mphungo, 75, 76, 77, 79
mzelelo, 88, 111
mzungu (pl. *wa-*), 15, 16

N
need/s, 2, 3, 12, 43, 44, 46, 47, 48, 49, 50, 53, 54, 56, 62n. 7, 63nn. 15, 16, 70, 74; child's, 55, 71, 74, 92n. 15, 99nn. 82, 83, 101, 103, 104, 106, 110, 111, 114, 116n. 5; nutritional, 35, 42, 43, 47, 48, 49, 53, 54, 57, 63n. 17, 66
newborn, 1, 21, 44, 45, 46, 47, 59, 62nn. 8, 12, 71, 72, 84, 92n. 12, 97n. 65, 98n. 69, 101, 102 103, 109; *see also* baby

nurturer/s, 2, 3, 4, 56, 58, 68, 71, 80, 82, 101, 102, 103, 104, 105, 106, 107, 108, 110, 111, 112, 113, 114; bad, 112; good, 102–105, 108, 112
nyumba, 26, 27, 31, 32–34, 35, 39n. 17, 40n. 19, 51, 69, 72, 73, 107
nyumba ya cilima, 27, 34
nyumba imbaha, 27, 34

O
offspring, 2, 33, 36, 42, 43, 63n. 22, 68, 69, 71, 85, 101, 108

P
placenta, 85, 101
post partum taboos, 4, 21, 53, 81–82, 83, 87, 92n. 9, 101
pregnancy/ies, 2, 3, 15, 37, 42, 45, 69, 70, 72, 82, 83, 84, 85, 86, 87, 91n. 1, 92n. 12, 109, 113, 114; early, 83; new, 51, 82–87, 89, 108; precocious, 87, 90, 108, 109; undesired, 83
pregnant, 63n. 23, 69, 82, 83, 84, 86, 90, 93n. 19, 98n. 75, 109, 111, 112, 113
prohibition/s, 17, 20, 37, 43, 68, 72, 80, 81, 83, 92n. 5, 95n. 58, 97n. 64, 100, 108
puberty, 16, 25n. 14, 36, 40n. 1, 42, 70, 96n. 58

R
relation/s, 12, 16, 20, 26, 33, 36, 43, 44, 68, 82; relationship/s, 10, 15, 16, 17, 19, 20, 22, 25nn. 14, 19, 28, 41, 44, 55, 56, 68, 71, 72, 81, 87, 88, 92n. 12, 95n. 40, 97n. 62, 99n. 80; extraconjugal 110 ; extramarital, 82, 88, 89, 90, 111; sexual, 16, 37, 70, 82, 87, 88, 97n. 64, 109; a mixture of sexual, 87–91
reproduction, 34, 92n. 2
reproductive: age, 21, 72; capacity, 101, 108, 113, 114, 116n. 16; cycle, 69; health, 2, 3; process, 85, 116n. 15; sphere, 113